PARADOXES IN
SCIENTIFIC
INFERENCE

PARADOXES IN SCIENTIFIC INFERENCE

Mark Chang

CRC Press
Taylor & Francis Group
Boca Raton   London   New York

CRC Press is an imprint of the
Taylor & Francis Group, an **informa** business

A CHAPMAN & HALL BOOK

CRC Press
Taylor & Francis Group
6000 Broken Sound Parkway NW, Suite 300
Boca Raton, FL 33487-2742

Version Date: 2012912

International Standard Book Number: 978-1-4665-0986-3 (Paperback)

Visit the Taylor & Francis Web site at
http://www.taylorandfrancis.com

and the CRC Press Web site at
http://www.crcpress.com

My paradoxical dedication:
To those who have chosen to read this book

Warning: Your brain may be reshaped unconsciously and involuntarily!

Preface

Learning and writing can be a quite enjoyable experience!

This book is about scientific inference and surrounding controversies. It presents a new way of learning about scientific inference through paradoxes. But what is a paradox? "Everyone is unique, just like everyone else." This statement is a simple example of paradox. A paradox is a statement or phenomenon that seems contradictory but in reality expresses a possible truth. It can be simply something that is counterintuitive. Paradoxes exist everywhere, in science, mathematics, philosophy, and in every corner of our lives.

Paradoxes are poems of science and philosophy that collectively allow us to address broad multidisciplinary issues intriguingly and profoundly within a tiny volume, such as this book. A true paradox is a concise expression that delivers a profound idea. A paradox provokes indisputably a wild and endless imagination. Paradoxes are absolutely a source of creativity. The study of paradoxes leads to ultimate clarity and at the same time evokes one's mind, indisputably.

The book analyzes paradoxes from many different perspectives: statistics, mathematics, philosophy, science, artificial intelligence, etc., elaborates findings, and reaches new and exciting conclusions.

Probability and Statistics provide an essential and yet powerful tool for scientific research because weighing scientific evidence involves uncertainties. Studying statistical paradoxes helps us understand how scientific evidence is measured together with surrounding uncertainties and controversies. On the other hand, studying scientific paradoxes helps us truly understand the different paradigms and controversies in statistics, and advances the theory of statistics in the right direction. That is the main reason I gather together many of the paradoxes in statistics and science in this single volume.

In additional to existing paradoxes from various sources, there are many paradoxes newly discovered during my study of this topic. Some are simple;

others are complicated. Although I have tried to keep mathematical formulations minimal, I have not totally eliminated them, so as to avoid mathematical anxiety that might result from either approach (I have marked those difficult sections with an asterisk). Such a limitation in mathematics will not impair me from having an in-depth discussion of the topics; instead, it allows a reader to focus on the conceptual challenges. Truly, in many places, it should challenge your knowledge, your intuition and conventional wisdom, and require you to make a necessary adjustment. Since many of the paradoxes are poems of science you will have different feelings, awakenings, sentiments, and understandings every time you read them.

It is not a book addressing the symmetric approaches in scientific inference (I don't even know if such a book exists), but a book outlining general principles and issues that everyone needs to know to be a rigorous and creative scientist. We will not focus on those most people know but on those that are often neglected. Although the book is written primarily for scientists and statisticians, many of the contents are also helpful for general audiences or those who have a college degree. With exercises at the end of each chapter, it can also be used as a textbook.

A good scientist is always critical and open-minded. He is always ready to face challenges and adjust his belief as new evidence arises. He always carefully examines new findings and their implications carefully, and separates out whenever it is possible a fact from belief, a plausible solution from the truth.

I always encourage everyone, especially those who do scientific research, to get training in statistics or applied statistics, not because of their mathematical aspect, but because of the ways they offer to approach problems.

As stated earlier, the approach taken in this book is unique: learning scientific inference through paradoxes. I will feel so rewarded if this book can help a younger reader in some way to become a thinker, a creative scientist, a rigorous statistician, or a sophisticated person living a simplified life.

Road Map
The book consists of five chapters, organized as follows:

Chapter 1, Joy of Paradoxes: A Random Walk
We will walk you "randomly" into the world of paradoxes. We will present you with many interesting examples, including simple paradoxical quotes for life and provocative paradoxes from social networks, the judicial system, and games. The paradox of identity switching, paradox of leadership, paradox of the murderer, paradox of the court, and the omnipotence paradox, while amusing, are stimulating and challenging.

We will further provide you with many motivating applications of paradoxes in accounting fraud detection, stock diversification, social economics, prediction of disease outbreak, the application of pesticides, machine maintenance, electric bell making, and internet package encryption.

We will share with you dozens of fascinating mathematical paradoxes and explore intriguing paradoxes involving probability and statistics. Most of them are count-intuitive examples that are commonly encountered in daily life.

Chapter 2, Mathematical and Plausible Reasoning
We will begin with the review of the two different notions of probabilities: Frequentist and Bayesian. We then unify them by examining the hidden assumption in conception of Bayesian probability—causal space. We conceptualize causal space using examples in daily life and unify the two paradigms.

We will introduce the very basics of formal logic before studying probability and probabilistic inference. To facilitate our discussion, we will utilize some well-known paradoxes, including the Boy or Girl Paradox, the Coupon Collector's Problem, Bertrand's Paradox, the Monty Hall Dilemma, the Tennis Game Paradox, the Paradox of Nontransitive Dice, the Paradox of Ruining Time, and the Paradox of Independence. We will not only provide fresh views on classical paradoxes but also present new paradoxes.

Chapter 3, Statistical Measures of Scientific Evidence
We advance probabilistic inference further to statistical inference. However, we will not focus on hands on statistical methodologies and expect you to perform statistical analyses. Instead, we will provide in-depth discussions and critiques on fundamental statistical principles and different statistical paradigms that are derived from these principles. Each of principles appears intuitive but the surrounding controversies are extremely complex. We will often use paradoxes in numerical form to effectively explain abstract concepts and complex issues.

We will describe the three statistical paradigms frequentism, Bayesianism, and likelihoodism, and the Decision Approach as well. After reviewing the concepts of statistical model, point estimate, confidence interval, p-value, type-I and type-II errors, and level of significance, we will provide a detailed discussion of five essential statistical principles and surrounding controversies: the conditionality principle, the sufficiency principle, the likelihood principle, the law of likelihood, and Bayes' law.

The discussion of controversies will be broadened to statistical analyses, including model fitting, data pooling, subgroup analyses, regression to the mean, and the issue of confounding. We put a large effort on the multiplicity issues due.

We will make an attempt to unify the statistical paradigms under the proposed principle of evidential totality and the new concept of causal space.

This chapter covers comprehensively and in depth most controversial issues in statistical inference. To make it effective, I will again borrow from the power of paradoxes.

Chapter 4, Scientific Principles and Inferences

We are going to study various principles, ideas, rules, methods, inferences, procedures, thought processes, intuitions, and controversies in scientific research. Some of the topics belong to scientific philosophy. We will use paradoxes to facilitate our discussion and make the abstract concepts tangible and easy to understand with minimum mathematical involvement.

We will discuss several fundamental and provoking questions in scientific philosophy: the definition of science and the meaning of understanding, theories of truth, and discovery versus invention. We give you a fresh view on the topic of Determinism versus Free Will. You will be surprised by how these simple terms with which we are so familiar can still cause so much confusion and controversy.

We will address the common issues in scientific research, share paradoxical stories about scientific research, and outline what the gold standard experiment should be. We will introduce a special kind of experiments that do not require a physical experiment and whereby no observations are needed. The thought experiments and examples including Galileo's leaning tower of Pisa discussion for free falling bodies, Maxwell's demon for a perpetual machine, and the Twin and Grandfather paradoxes in regard to time travel.

Logical or axiom systems are usually considered the most rigorous schemes in science. Surprisingly, Gödel's Incompleteness theorem proved that an axiom system is, necessarily, either incomplete or inconsistent. Fitch proved that if all truths are knowable in principle then all truths are in fact known. Probably no scientific law is simpler than the pigeonhole principle. However, very strange things happen when a human brain is considered as a set of pigeonholes. We will discuss many more paradoxes like these.

Game theory has great applications in economics and social science. We will stimulate our discussion with four different paradoxes in game theory.

Chapter 5, Artificial Intelligence

This chapter has two distinct parts, the paradox in artificial intelligent (AI) and the architecture for new AI agent. The later part is to provide a way to prove the "human-race" AI agent. Those who are not interested in AI building or do not have a computer science or AI background may choose to skip the second part.

The idea of AI was proposed by Alan Turing in the Turing Machine. There are great debates on whether any computer can be built to perform like a human. I asset this can simply be a definitional problem, as illustrated by the "Ship of Theseus". The possibility of building "machine-race" humans is discussed through paradoxes of emotion, understanding, and discovery. The possibility is further illustrated by the paradox of dreams, the paradox of meta-cognition, and recent developments in swarm intelligence.

There are several fundamental differences between the proposed AI approach and the existing AI approaches: (1) The new AI has very limited built-in initial concepts (only 14) —virtually an initial empty brain, whereas traditional AI has a built-in knowledge database and/or complex algorithms. (2) The new AI is language-independent, whereas traditional AI is language-dependent (English and Chinese AIs have different architecture). (3) Unlike traditional AI, the new AI is based on the recursion of simple rules over multiple levels in learning. (4) The new AI's learning largely depends on progressive teaching: It can learn very broadly about many things but it is slow, whereas traditional methods have a fast learning speed but a narrow learning scope.

Fascinating, Exotic, Provocative, Enlightening! Indeed, I was excited during the whole process of, as well as after, writing this book. So I hope you will be too, when you are reading it.

My last note to readers: If you are an advanced reader you should read everything; if you are an intermediate reader, you can skip Chapter 3 and the second part of Chapter 5; if you are a naïve/quick reader, you might just read Chapters 1 and 4, and the first part of Chapter 5.

Acknowledgment

My work is built on the previous achievements of and support from many others. I'd like to express my sincere thanks to Dr. Robert Pierce for his comprehensive review and numerous suggestions. I'd also like to give thanks to the anonymous reviewers. Their encouragement and constructive comments have made a great impact on the final outcome of my work—delivering intricate thoughts without unnecessary mathematical sophistication. There are a vast number of books, research papers, and online publications that have served as excellent resources for me in accomplishing this book. I earnestly thank their authors for their outstanding work. Particularly, I'd like to mention two books that have inspired me: "Paradoxes in Probability Theory and Mathematical Statistics" by Professor Gabor Székely, and "Chance, The Life of Games & The Game of Lif" by Professor Marques de Sá. Others have been cited within the text and listed as references. Special thanks to Ms. Helena

Zhang for drawing the illustrations for me. Last but not least, I'd like to thank Mr. David Grubbs from Taylor and Francis for providing me the opportunity to work with him on this book project.

Mark Chang (張揚)

Contents

Chapter 1

The Joy of Paradoxes: A Random Walk

Quick Study Guide

In this opening chapter, we will walk you "randomly" into the world of paradoxes. We will present many interesting examples, including simple paradoxical quotes about life and provocative paradoxes from social networks, the judicial system, and games. The paradox of identity switching, paradox of leadership, paradox of the murderer, paradox of the court, and the omnipotence paradox, while amusing, are also stimulating and challenging.

We will further provide you with many motivating applications of paradoxes in accounting fraud detection, stock diversification, social economics, prediction of disease outbreak, the application of pesticides, machine maintenance, electric bell making, and Internet package encryption.

We will share with you dozens of fascinating mathematical paradoxes and explore intriguing paradoxes involving probability and statistics. Most of them are counterintuitive examples that are commonly encountered in daily life.

1.1 Introduction to Paradox

A paradox is a statement or proposition that seems contradictory but in reality expresses a possible truth. A paradox can be simply something that is counterintuitive. It can also be a phenomenon at variance with normal ideas of what is probable, natural, or possible. A simple example would be the paradox of thrift.

The paradox of thrift can be stated like this: Suppose a large group of people decide to save more. One might think that this would necessarily mean a rise in national savings. However, if falling consumption causes the economy to fall into a recession, incomes will fall, and so will savings, other

1

things being equal. This induced fall in savings can largely or completely offset the initial rise. The paradox is, narrowly speaking, that the total savings may fall even when individual savings attempt to rise, and, broadly speaking, that increases in savings may be harmful to an economy.

Paradoxes have played an important role in all branches of science, mathematics (statistics), and engineering since they can bring clarity to concepts, lead to the creation of new ideas, and to new approaches to solving problems. Paradoxes can sharpen one's mind and help deepen one's understanding of a problem. Indeed, paradoxes are a source of innovation and creativity. A paradox is an immediate attraction to any great researcher. As the philosopher Søren Kierkegaard once put it: "The paradox is the source of the thinker's passion, and the thinker without a paradox is like a lover without feeling."

This book is a study of the nature of scientific inference through controversies or, more precisely, paradoxes. While recognizing that "a classic is something that everybody wants to have read and nobody wants to read"— Mark Twain (Székely, 1986), I have carefully selected a set of classic paradoxes that can enormously stimulate your brain ... with amusement.

1.1.1   *Paradoxes about Life*

Paradoxes can teach helpful life lessons. Here are some simple examples.

A man who fears suffering is already suffering from what he fears.
                                                          —**Montaigne**
Happiness is the absence of striving for happiness.
                                                          —**Chang-Tzu**
Reach the goal by giving up the attempt to reach it.
                                                          —**Shaw**
If you want what you get, you will get what you want.
                                                          —**Shaw**
Failure is the foundation of success, and success the lurking place of failure.
                                                          —**Lao-Tzu**
The more they give, the greater their abundance.
                                                          —**Lao-Tzu**
Shared joy is a double joy; shared sorrow is half a sorrow.
                                                          —**Swedish Proverb**
Art is a lie that makes us realize the truth.
                                                          —**Picasso**
I shut my eyes in order to see.
                                                          —**Gauguin**

A thing rests by changing.

—Heraclitus

Painting is easy when you don't know how, but very difficult when you do.

—Degas

The notes I handle no better than many pianists. But the pauses between the notes, ah, that is where the art resides.

—Schnabel

If you wish to preserve your secret, wrap it up in frankness.

—Smith

If infinities can be stored in and handled by a finite human brain, then what are the infinities really?

—Mark Chang

Probability is a measure of possibility, but probability 0 does not mean impossibility.

—Mark Chang

I keep my personal notes of paradoxes. Here I cited one: "There is no absolute truth; everything is changing." But how about the statement itself. If it's true it has invalidated the statement; if it's not true, then it implies there is an absolute truth. Do you think we just proved there is "absolute truth"?

You may have heard the popular song by Leo Sayer: "I love you more than I can say." This statement is clearly a paradox that appears to be true. The fact is if the statement is true, then your love is no more than what you just said. If the statement is false, it also implies your love is no more than what you can say. Either way the statement "I love you more than I can say" cannot be true.

Let's look at the next one. Is it really immoral to break your promise? Maybe not, if you have not promised to keep the promise. Moving on, thinking can be difficult. To think about thinking is challenging, but to think about thinking about thinking is beyond imagination. I wonder if anyone has confidence in Aristotle's logic at those high levels of recursion? Here is probably one of the most amazing features of our brain: A finite brain is a place for the concept of infinity. If infinities can be stored in and handled by a finite human brain, then what are the infinities really?

The amusement of paradoxes does not stop there. The paradox of blackmail goes like this: There is nothing illegal about asking someone for money, nor is it illegal to threaten to report someone's theft. However, if you threaten to expose someone's crime unless she pays you money, you are guilty of blackmail (Clark, 2007, p.25). Even more amazingly, can you image that the ratio of a smaller number to a greater one can be equal to the ratio of a greater number to a smaller one? If you can't, here is an example, $(-1) : 1 = 1 : (-1)$.

I can very easily create a few more just for your entertainment:

Paradox of Democracy and Dictatorship

A country can have national democracy but at the same time practice international dictatorship. A country can have national dictatorship while promoting international democracy. If one dictatorially makes a change towards democracy, what should it be called, dictatorship or democracy?

Paradox of Identity Switching

Prof. Lee is getting old. He expressed his wish to have a younger and healthy body, whereas a healthy young student, John, truly admires Prof. Lee's knowledge. After they learned each other's wishes, they decided to switch their bodies or knowledge, however you wish to say it. In the operation room, the professor's knowledge (information) was removed from his brain and transferred to John's brain. At the same time, whatever original information that was in John's brain was removed and transferred to Doctor Lee's brain. The operation was carried out using the "incredible machine." Now the question is: Who is who after the operation? Would they both be happy with the operation? Please think, and then think again before you answer.

Paradox of Leadership

It seems to be only common sense that following a good leader is a smart thing to do. However, this is not always true, as illustrated by the following amazing paradox that I modified and extended from the example given by Székely (1986, p. 171). Suppose that A, B, C, D, and E are the five members of a jury in a trial. Guilt or innocence for the prisoner is determined by simple majority rule. There is a 5% chance that A gives the wrong verdict, for B, C, and D it is 10%, and E is mistaken with a probability of 20%. When the five members vote independently, the probability of bringing the wrong verdict is about 1%. Paradoxically, this probability increases to 1.5% if E (who is most probably mistaken) abandons his own judgment and always votes the same as A (who is most rarely mistaken). Even more surprisingly, if the four members B, C, D, and E all follow A's vote, then the probability of delivering the wrong verdict is 5%, five times more than that when they vote independently. The paradox implies it is better to have your own opinion even if it is not as good as the leader's opinion, in general. This is, in fact, a great feature of collective intelligence, as further discussed in Chapter 4.

Simpson's Paradox

We are all concerned about our health and are cautious about drug alternatives when we are sick. Suppose two drugs, A and B, are available for a disease. The treatment effect (in terms of response rate) is 40/110 for B, better than 35/110 for treatment A. Thus, we will prefer treatment B to A. However, after further looking into the data for males and females separately, we found that the treatment effect in males is 5/10 with A, better than 38/100 with B, and treatment effect in females is 30/100 with A, better than 2/10 with B. Therefore, whether female or male, we will prefer treatment A to B. Should we take treatment A or B? You will see more discussions of this topic in Chapter 3.

	Drug A	Drug B
Male	5/10	38/100
Female	30/100	2/10
Total	35/110	40/110

Paradox of the Murderer

Raymond Smullyan (1976) discovered the following interesting paradox: A caravan of three (A, B, and C) were going through the Sahara desert, where one night they pitched tents. A hated C and decided to murder him by putting poison in the water of his canteen (C's only water supply). B, who also wanted to murder C, didn't realize that C's water was already poisoned, and he drilled a tiny hole in C's canteen so that the water would slowly leak out. As a result, several days later C died of thirst. The question is, who was the murderer, A or B? According to one argument, B was the murderer, since C never did drink the poison put in by A; hence, he would have died even if A hadn't poisoned the water. According to the opposite argument, A was the real murderer, since B's actions had absolutely no effect on the outcome; once A poisoned the water, C was doomed, and hence A would have died even if B had not drilled the hole.

Which argument do you agree with?

The Knowing-The-Fishing Bridge

The Garden of Harmonious Interest in Beijing, China, known as "the Garden with a Garden," is basically a garden in a waterscape. Spanning the vast expanse of the lake and pools are five bridges that used to be Bridge Hao, each quite different from the others. The most famous of them is the bridge known as the Knowing-the-Fishing Bridge. It is said that more than 2,500 years ago during the Warring States Period, two philosophers named Zhuang Zi and Hui Zi had an amusing argument while strolling on Bridge Hao:

Zhuangzi said: *Look how happy the fish are!*
Huizi asked: *You are not a fish, how do you know they are happy?*
Zhuangzi answered: *You are not I, how do you know I don't know?*
Huizi added: *I am not you, so, I don't know you. Likewise, you are not fish,
    so you don't know if they are happy or not!*

The question now is: Did Zhuangzi know that fish are happy or not? Should we try respecting every living being, not just human beings, from their point of view? Or is that even possible (Section 4.1.2)?

Paradox of "What Is Not"

The *Tao Te Ching* is a great treasure house of wisdom. Written as early as the sixth century B.C. and composed of only 5000 Chinese characters, it has become one of the classic works of spiritual enlightenment. The Tao offers a much-needed alternative to our fragmented, modern ways of thinking, feeling, and behaving. To live life in accordance with "Tao" is to be in harmony with others, with the environment, and with oneself (Dale, 2002). During the past 2400 years, it is estimated that there have been about 1400 interpretations, with perhaps 700 extant and over 100 in English. The paradox of "What is not" is an example of Dale's interpretation (Dale, 2002):

We join thirty spokes to the hub of a wheel,
yet it's the center hole that drives the chariot.

We shape clay to birth a vessel,
yet it's the hollow within that makes it useful.

We chisel doors and windows to construct a room,
yet it's the inner space that makes it livable.

Thus do we create what is to use what is not.

1.1.2   *Paradoxes of Self-Reference*

Self-reference occurs when a statement refers to itself, thus sometimes creating a paradoxical situation. Here are some simple examples.

OMNIPOTENCE PARADOX

CAN GOD CREATE A STONE HE CAN'T LIFT?

Omnipotence Paradox

The Omnipotence Paradox generally states that if a being can perform any action, then it should be able to create a task it is unable to perform, and hence, it cannot perform all actions. Yet, on the other hand, if it cannot create a task it is unable to perform, then something it cannot do exists (wikipedia.com). For example, "Can an omnipotent being create a rock so heavy that it cannot lift it?" This question generates a dilemma. The being can either create a stone which it cannot lift, or it cannot create a stone which it cannot lift. If the being can create a stone that it cannot lift, then it seems that it ceases to be omnipotent. If the being cannot create a stone which it cannot lift, then it seems it is already not omnipotent.

Paradox of the Court

The Paradox of the Court is a very old problem in logic from ancient Greece. There are several versions of the paradox. Here is one you can find at wikipedia.org: It is said that the famous sophist Protagoras took on a student, Euathlus, on the understanding that the student would pay Protagoras for his instruction after he had won his first case. Protagoras demanded his money as soon as Euathlus completed his education. At the court, Protagoras argued that if he won the case he would be paid his money. If Euathlus won the case, Protagoras would still be paid according to the original contract, because Euathlus would have won his first case.

Euathlus, however, claimed that if he won, then by the court's decision he would not have to pay Protagoras. If on the other hand, Protagoras won then Euathlus would still not have won a case and therefore would not be obliged to pay according to the original contract.

The question is: Which of the two men is right?

1.1.3   *Six Degree of Separation*

Six degrees of separation refers to the idea that everyone is on average approximately only six steps away from any other person on Earth. It was probably coined by Frigyes Karinthy and popularized by a play written by John Guare.

In the 1960s, a social psychologist Stanley Milgram found that he could send a letter to a random person in Nebraska or Boston and have it reach a random target person in Massachusetts. Milgram found that on average it would take 5.2 intermediaries for his letter to go from the first person to its destination, via each person's social network. Milgram has sent the optimistic message that each of us is only a few social steps away from everyone else in the world. We really do live in a small world

However, Milgram's work has been challenged by Professor Judith Kleinfeld, who found Milgram had used relatively prominent people in society as the starting and ending points for his chain letter; plus, a very high percentage of letters never arrived at their destination at all, and many don't test links across class or geographical boundaries (www.spring.org.uk).

In 2001, Duncan Watts, a professor at Columbia University, attempted to recreate Milgram's experiment on the Internet, using an e-mail message as the "package" that needed to be delivered, with 48,000 senders and 19 targets (in 157 countries). Watts found that the average number of intermediaries was around six (www.pleacher.com). In 2003, though, some support for Milgram's idea was found by Watts and colleagues at Columbia University in a paper published in *Science*. E-mailers were asked to try and forward a message to one of 18 target people in 13 different countries, going via friends and acquaintances. All together more than 60,000 people took part. They found that successful chains were completed in between 5 and 7 steps, similar to Milgram's results. A 2007 study by Jure Leskovec and Eric Horvitz examined a dataset of instant messages composed of 30 billion conversations among 240 million people. They found that the average path length among Microsoft Messenger users was 6.6. Unfortunately, such e-mail replication had the same problem as Milgram's original study: Many chains simply broke down. Also, like Milgram's study, the targets were relatively visible in society: One was a vet, another a policeman, and another a technology consultant, but no factory operatives or convenience store workers existed among the 18 targets in this sample (www.spring.org.uk/2008/08/).

I wonder if the letter or "message" was taking the shortest path to reach each destination. Besides, the degree of separation is dependent on the definition of connection. For instance, we can define a connection if two people have the same father or use the same website or have an account on the same website (such as Facebook or Linkedin). In the latter cases, many people who use the Internet have only two degrees of separation. What really surprises me is that the world before the information age was much smaller than we thought, and after the information age it is naturally even smaller.

Small-world networks have been widely studied recently. In a small-world network, the number of people (nodes) is not necessarily small, but the average diameter $\bar{d}$ (average number of steps connecting two nodes) of the network increases slowly as the total number of nodes N increases. Mathematically, it increases on the order of $\ln N$: In layman's terms, the average steps needed to connect two people in a small-world network is proportional to $\ln N$ for sufficiently large N. Equivalently, we can say that the number of nodes in a small-world network increases exponentially with $\bar{d}$; i.e., $N \sim \exp\left(\bar{d}/d_0\right)$, where d_0 is the "characteristic length" of the network. Network theory is mainly built on graph theory and game theory. Graph theory is the study of network structure, while game theory provides models of individual behavior in settings where outcomes depend on the behavior of others (Easley and Kleinberg, 2010, p. 7).

1.1.4 *Paradox of Music and Noise*

I am amused, actually more surprised, by the fact that only 13 different musical scales (others are multipliers of these 13 scales) are needed to make virtually any music in the world. Why are the intervals between the music scales so universal, but paradoxically, the languages are so different? There must be important and very interesting intrinsic properties that we (at least I) have not been able to recognize yet. On the other hand, noise can be any sound that does not follow the scales.

The sine wave or sinusoid is a mathematical function that describes a smooth repetitive oscillation. It occurs often in pure mathematics, as well as physics, signal processing, electrical engineering, and many other fields. Its most basic form as a function of time (t) is

$$Y = A \sin(\omega t),$$

where Y is, e.g., the displacement; A, the amplitude, the peak deviation of the function from its center position; and ω, the angular frequency, which specifies how many oscillations occur in a unit time interval, in radians per second.

If the angular frequency ω is uniformly distributed, then the amplitudes of the sum of sinusoids are normally distributed. Such a random process is called white noise. Conversely, Newton's decomposition of white light reproducing the rainbow phenomenon is a well-known experiment. The French mathematician Joseph Fourier introduced the so-called Fourier transform based on a sum of sinusoids, which became a widely used tool for analysis in many areas of science and engineering. The Fourier transform of a probability distribution is the so-called characteristic function, a powerful tool in statistics.

White noise sounds like streaming water. The spectrum of violin music is similar to a sum of sinusoids. The fluctuations of Brownian motion are correlated in a given time scale and the corresponding (autocorrelation) spectrum varies as $1/\omega^2$. Brownian "music" sounds less chaotic than white noise and when played at different speeds, sounds different while nevertheless retaining its Brownian characteristic.

According to Marques (2008), sequences of fractional Brownian motion (a random process with memory) exhibit correlations and their spectra vary as $1/\omega^a$, where $a = 1 + 2H$, taking values between 1 and 3 ($H = 0.5$ corresponding to Brownian motion). Music by Bach has a $1/\omega$ spectrum. Brownian music with $1/\omega^2$ spectrum gives one a feeling of monotony, whereas "noise" with $1/\omega$ spectrum gives some aesthetic feeling. Marques (2008) pointed out that there are many natural phenomena exhibiting scale invariance, known as $1/\omega$ phenomena. For instance, an animal's sense is measured by a $1/\omega$ or $\log \omega$ scale. There are also many socioeconomic $1/\omega$ phenomena, such as the wealth distribution in many societies, the frequency with which words are used, and the production of scientific papers.

1.1.5 *Paradox of a Fair Game*

Not all the games that appear to be fair are, in fact, fair. Marques de Sá, (2008, p. 62) describes the following remarkable game.

Imagine the following game between two opponents A and B. Each player constructs a stack of one-dollar coins. To play the game, each takes out a coin from the top of his stack at the same time. The winning rules are defined as follows:

(1) If head-head appears, A wins \$9 from B.
(2) If tail-tail appears, A wins \$1 from B.
(3) If head-tail or tail-head appears, B wins \$5 from A.

This game, appearing to be a fair, is actually an unfair one that allows player B win \$0.8 per round on average. Player A will tend to pile the coins up in such a way as to exhibit heads more often, expecting to find many heads-heads situations. On the other hand, B, knowing that A will try to show up heads, will try to show tails. One might think that, since the average gain in the situation that A wins, i.e., $(9+1)/2$, and in the situation that B wins, i.e., $(5+5)/2$, are the same, it will not really matter how the coins are piled up as long as the opponent does not know your stack.

However, this is not true. Suppose A piles up the coins so that they always show up heads. Let t represent the fraction of the stack in which B puts heads facing up. The gain of player B is given by

$$G_H = -9t + 5(1-t) = -14t + 5.$$

If the coins of stack A always show tails, one likewise obtains

$$G_T = 5t - 1(1-t) = 6t - 1.$$

Letting $G_H = G_T$, we can obtain $t = 0.3$ and B gains $G_H = G_T = \$0.8$. This result also applies to any stack segment where player A always shows heads (tails) up. In other words, the result is also applicable to the situation when A has a mixed head-tail stack. This means that if player B stacks his coins in such a way that three-tenths of them, randomly distributed in the stack, show up heads, he will expect to win, during a long sequence of rounds, an average of \$0.8 per round.

1.1.6 *DNA Paradox in Justice System*

"Is DNA technology the ultimate diviner of guilt or the ultimate threat to civil liberties? Over the past decade, DNA has been used to exonerate hundreds and to convict thousands. Its expanded use over the coming decade promises to recalibrate significantly the balance between collective societal interests (security) and individual freedom." Lazer (2004)

"On May 1, 1990, Roy Criner was convicted of the 1986 murder and rape of Deanna Ogg and sentenced to ninety-nine years in prison. Seven years later, a DNA test conducted on the semen left on Ogg's body excluded Criner as its source. However, a state district court decision to grant Criner a new trial was overturned in a five-to-four decision by the Texas Court of Criminal Appeals, which offered the theory that Ogg might have had consensual intercourse prior to being raped by Criner and that Criner might have worn a condom or not ejaculated. Three years later, a cigarette butt from the crime scene was tested: It contained DNA from Ogg and the individual who was the source of the semen on Ogg's body—not Criner, thus undermining the appeals court's theory." Lazer (2004)

The first use of DNA in a criminal investigation involved a "DNA dragnet": to capture the aptly named Colin Pitchfork in 1987. DNA technology is commonly used in forensic science. Since then, the over 140 convict exonerations in the United States that have resulted from postconviction DNA analysis raise two serious questions about the U.S. criminal justice system. The first question is whether the system is receptive to evidence that it has erred in a particular case. The second is whether these exonerations highlight systematic fault lines in the system itself.

Since DNA database laws authorizing the creation of DNA databases rapidly spread through all 50 states during the 1990s, all states in the United States have passed laws creating DNA databases—one in each state for DNA profiles of individuals convicted (and sometimes, just arrested) for particular crimes, and one for profiles developed from DNA evidence left at the scenes of crimes.

Like many other cases, the O.J. Simpson case was an issue of trust: not trust in DNA technology, but trust in the system. Did the Los Angeles Police Department handle the samples from Simpson properly? Might they have engaged in a conspiracy to frame Simpson? In the end, the technology was not a cure for distrust in the system. The system could be wrongly convicting individuals because of racial biases, the underfunding of the defense, and unreliable police practices. DNA technology can sometimes offer an ex post correction of these errors; however, the technology cannot be trusted in the hands of the system as the system is currently constructed (Lazer, 2004).

The notion of using DNA technology in justice systems is that DNA matching technology is recognized as a very powerful tool for conviction in court since the probability of DNA matching is rare, but the intended or unintended errors in experiments (of DNA) are not rare at all.

1.2 Applications of Paradoxes

1.2.1 *Benford's Law and Detecting Accounting Fraud*

Benford's law, also called the first-digit law, is counterintuitive and surprising to many. It says that in lists of numbers from many real-life sources of data (e.g., electricity bills, street addresses, stock prices, population numbers, death rates, lengths of rivers, physical and mathematical constants), the leading digit is, surprisingly, not distributed uniformly, and the distribution is independent of the units chosen. For instance, data from the Political Calculations (2009) show that distribution of the digits are 30.1%, 17.6%, 12.5%, 9.7%, 7.9%, 6.7%, 5.8%, 5.1%, and 4.6% for digits 1 to 9, respectively. The heights of the 60 tallest structures in the world by category shows the distribution of the digits being 43.3%, 11.7%, 15.0%, 10.0%, 6.7%, 1.7%, 3.3%, 8.3%, and 0.0% when measured in meters and 30.1%, 17.6%, 12.5%, 9.7%, 7.9%, 6.7%, 5.8%, 5.1%, and 4.6% when measured in feet. Benford's law states precisely that the leading digit n ($n = 1, 2, \ldots, b-1$) in base b ($b > 2$) occurs with probability (Benford, 1938; Gorroochurn, 2012, p. 233—239)

$$P(n) = \log_b \left(1 + \frac{1}{n}\right).$$

The limitation $b > 2$ excludes the trivial cases of binary systems ($b = 2$), in which all numbers begin with 1, and unary systems ($b = 1$). The infinite integer sequences such as the Fibonacci numbers and the factorials provably satisfy Benford's law exactly (wikipedia.org).

The physicist Frank Benford stated the paradox in 1938. However, it had been previously noticed by Simon Newcomb in 1881. The law can alternatively be explained by the fact that, if it is indeed true that the first digits have a particular distribution, it must be independent of the measuring units used. This means that if one converts from say feet to meters, the distribution must be unchanged—it is scale invariant—and the only distribution that fits this is one whose logarithm is uniformly distributed.

Benford's law has been used in fraud detection. As early as 1972, Hal Varian suggested that the law could be used to detect possible fraud in lists of socio-economic data submitted in support of public planning decisions. Based on the plausible assumption that people who make up figures tend to distribute their digits fairly uniformly, a simple comparison of first-digit frequency distribution from the data with the expected distribution from Benford's law ought to highlight any anomalous results. Following this idea, Mark Nigrini (1999) showed that Benford's law could be used as an indicator of accounting and expense fraud. In the United States, evidence based on Benford's law is legally admissible in criminal cases at the federal, state, and local levels. Benford's law has been invoked as evidence of fraud in the 2009 Iranian elections (Battersby, 2009). However, other experts consider Benford's law essentially useless as a statistical indicator of election fraud in general (Deckert et al., 2010) (wikipedia.org).

1.2.2   *Parrondo's Paradox and Stock Diversification*

In game theory, Parrondo's Paradox can be stated as follows: Suppose there are two games, each with a higher probability of losing than winning. It is possible to construct a winning strategy by playing the games alternately. The paradox is named after its creator Juan Parrondo, a Spanish physicist, in 1996. The paradox was inspired by the mechanical properties of ratchets, the familiar saw-tooth tools used in automobile jacks and in self-winding watches.

An example of Parrondo's Paradox is coin tossing in gambling. Consider playing two games, Game A and Game B, with the following rules. Denote C_t as our capital at time t immediately before we play a game. Winning a game earns us \$1, and losing requires us to surrender \$1. It follows that $C_{t+1} = C_t + 1$ if we win at step t, and $C_{t+1} = C_t - 1$ if we lose at step t.

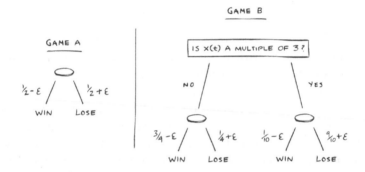

In Game A, we toss a biased coin, Coin 1, with probability of winning $P_1 = (1/2) - \varepsilon$. If $\varepsilon > 0$, this is clearly a losing game in the long run.

In Game B, we first determine if our capital is a multiple of some integer M. If it is, we toss a biased coin, Coin 2, with probability of winning $P_2 = 1/10 - \varepsilon$. If it is not, we toss another biased coin, Coin 3, with probability of winning $P_3 = 3/4 - \varepsilon$. The role of modulo M provides the periodicity as in the ratchet teeth.

It is clear that by playing Game A we will almost surely lose in the long run. Game B is a Markov chain, and an analysis of its state transition matrix (with modules $M = 3$) shows that the steady state probability of using Coin 2 is 0.3836, and that of using Coin 3 is 0.6164 (Minor, 2003). Therefore, the expected gain in game B is $0.3836(1/10 - \varepsilon) + 0.6164(3/4 - \varepsilon) = 0.500\,66 - \varepsilon$. Thus, B is a losing game if $\varepsilon > 0.0007$.

However, when these two losing games are played in some alternating sequences; e.g., two games of A followed by two games of B ($AABBAABB....$), the combination of the two games is, paradoxically, a winning game. One game of A followed by two games of B ($ABBABB....$) is also a winning game. However, $ABABAB...$ is a losing game. When $\varepsilon=0.05$ and \$1 bets, game B is a losing game ($-\$0.01$ per game), whereas the gain per game for the three sequences of the games are approximately \$0.02, \$0.01, and $-\$0.01$, respectively. The apparent paradox has been explained using a number of sophisticated approaches, including Markov chains (Harmer and Abbott, 1999a), flashing ratchets (Harmer and Abbott, 1999b), and Simulated Annealing (Harmer, et al., 2000). One way to explain the apparent paradox is as follows.

While Game B is a losing game under the probability distribution that results for C_t modulo M when it is played individually (C_t modulo M is the remainder when C_t is divided M), it can be a winning game under other distributions, as there is at least one state in which its expectation is positive. The value M determines the dependence between Games A and B, so that a player is more likely to enter states in which Game B has a positive expectation, allowing it to overcome the losses from Game A.

The Parrondo Paradox has been widely studied in financial marketing. For instance, Key, Klosek, and Abbott (2006) study the paradox to incorporate alternative switching rules, including the possibility of switching plays based on the result of a coin toss. Stutzer (2010) demonstrates that a Parrondo Paradox may arise in a simple model of Blackjack-type betting. The required switching rules are also in accord with reality, which are the two most common ways of implementing the oft-advocated principle of diversification. The method is outlined as follows.

The bettor chooses a constant fraction f of his table fortune to bet on each (Bernoulli) play, having a gross return per dollar invested equal to $1+f$ with probability $\pi > 1/2$, or $1-f$ with probability $1-\pi$. It is assumed that the bet is favorable, i.e., $\pi > 1/2$, so the bet has a positive "edge" $(2\pi-1)$ and positive expected net return $(2\pi - 1)f$ per play. Stutzer shows through simulations that for $\pi = .51$ and $f = 5\%$, repeated play of the single game resulted in a median loss of around 22% of the initial stake, while merely splitting the initial stake between two such identical games resulted in a median gain of around 20%! Here is why.

The total fortune after n plays of the Bernoulli trials is

$$F_n = F_0 \left[(1 + f)^w (1 - f)^{w-n} \right],$$

where F_0 is the bettor's initial value and w is the number of wins. For $\pi = .51$ and $f = 5\%$, the computer simulation results (based on n Bernoulli trials or a binomially distributed w) show a median $F_n = 0.778F_0$.

Now suppose the bettor plays two identical games alternately. He diversifies, placing half his funds in each game. He then alternates play between the two, and lets the funds "ride" in each game. His total fortune after n plays of both games (determined by $2n$ Bernoulli trials with w_1 and w_2 being the number of wins in the two games) is

$$F_n = \frac{1}{2}F_0 \left[(1+f)^{w_1}(1-f)^{w_1-n}\right] + \frac{1}{2}F_0 \left[(1+f)^{w_2}(1-f)^{w_2-n}\right]$$

Again, for $\pi = .51$ and $f = 5\%$, the computer simulation results show that the median of F_n is $1.2F_0$.

Remember, winning in the median indicates that most people will win if they play the same game with the strategy, whereas winning in the mean presents an expected win for a single bettor when playing the game repeatedly.

The interesting questions are: Can you use the Paradox to diversify your stock to become a winner when the economy is in bad shape? How will the time dependency of π affect your strategy? Finally, if everyone applies the same strategy and wins, who will lose? Is this a strategy causing many smaller losers and a fewer larger winners?

1.2.3   *Arrow's Paradox in Social Economics*

In social choice theory, Arrow's impossibility theorem (Arrow's Paradox) is named after economist Kenneth Arrow (Easley and Kleinberg, 2010, p. 657), who demonstrated the theorem in the original paper "A Difficulty in the Concept of Social Welfare." One can say that the contemporary paradigm of social choice theory started from this theorem. Arrow was a corecipient of the 1972 Nobel Memorial Prize in Economics.

The theorem proves that no voting system can be designed that satisfies these three "fairness" criteria:

(1) If every voter prefers alternative A over alternative B, then the group prefers A over B.
(2) If every voter's preference between A and B remains unchanged, then the group's preference between A and B will also remain unchanged (even if voters' preferences between other pairs like A and C, B and C, or C and D change).
(3) There is no "dictator": No single voter possesses the power to always determine the group's preference.

We might consider Arrow's theorem to be a paradox because it seems to violate our intuitive sense of what an election can and should be. Arrow's theorem is a mathematical result, but it is often expressed in a nonmathematical way by a statement such as "No voting method is fair," "Every ranked voting method is flawed," or "The only voting method that isn't flawed is a dictatorship." These simplifications of Arrow's result should be understood in context. What Arrow has proved is that any social choice system respecting universality, unanimity, and independence of irrelevant alternatives (IIA) is a dictatorship. The terms, universality, unanimity, and IIA are defined as follows.

Universality: For any set of individual voter preferences, the social welfare function should yield a unique and complete ranking of societal choices.

Unanimity (Pareto efficiency): If every individual prefers a certain option to another, then this must also be the resulting societal preference.

IIA: The social preference between A and B should depend only on the individual preferences between A and B (Pairwise Independence). More generally, changes in individuals' rankings of irrelevant alternatives (ones outside a certain subset) should have no impact on the societal ranking of the subset.

Various theorists have suggested weakening the IIA criterion as a way out of the paradox. Proponents of ranked voting methods contend that the IIA is an unreasonably strong criterion. Advocates of this position point out that failure of the standard IIA criterion is trivially implied by the possibility of cyclic preferences (wikipedia.org). Let's see the following example of ballots, where $\succeq$ denotes preference:

- 1 vote for $A \succeq B \succeq C$
- 1 vote for $B \succeq C \succeq A$
- 1 vote for $C \succeq A \succeq B$

If these votes are cast, then the pairwise majority preference of the group is that A wins over B, B wins over C, and C wins over A: These yield rock-paper-scissors preferences for any pairwise comparison. In this circumstance, any aggregation rule that satisfies the very basic majoritarian requirement that a candidate who receives a majority of votes must win the election will fail the IIA criterion if social preference is required to be transitive (or acyclic). To see this, suppose that such a rule satisfies IIA. Since majority preferences are respected, the society prefers A to B (two votes for $A \succeq B$ and one for $B \succeq A$), B to C, and C to A. Thus, a cycle is generated, which contradicts the assumption that social preference is transitive (wikipedia.org).

So, what Arrow's theorem really shows is that voting is a nontrivial game, and that game theory should be used to predict the outcome of most voting mechanisms. This could be seen as a discouraging result, because a game need not have efficient equilibria; e.g., a ballot could result in an alternative nobody really wanted in the first place, yet everybody voted for.

1.2.4 *Friendship Paradox and Disease Outbreak Prediction*

Would you be surprised if you are told by the sociologist Scott Feld that the friends of any given individual are likely more popular than the individual herself? The phenomenon of people finding whose friends have more friends than they do can be partially understood by recognizing the difference between the distribution of numbers of friends of individuals and the distribution of the numbers of friends of friends. The distribution of friends of individuals is just the usual distribution of the numbers of friends that we would usually examine, but the distribution of friends of friends includes some of the same individuals over and over (Feld, 1991). For example, in the network of five people shown in the following figure, John has 3 friends, and they (Joe, Bob, and Frank) have 4, 2, and 2 friends, respectively. The average number of friends of friends is equal to the total number of friends of friends (34) divided by the total number of friends (12). See the summary of calculations in the following table.

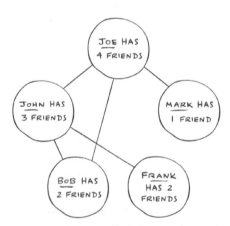

SOCIAL NETWORK: FRIENDSHIP PARADOX

The calculations in the example indicate that a higher proportion of individuals will be below the mean number of friends of friends than below the mean number of friends of individuals. However, there can be exceptions if we aggregate the social network in a special way.

Mathematically, the relations between the mean number u of friends of a random person and mean number of friends U that a typical friend has can be expressed as $U = u + \sigma/u$, where σ is the standard deviation of the numbers of friends that a friend has. Strictly speaking, the comparisons should be based on the median, not mean of the number of friends.

Feld (1991) pointed out that the tendency for individuals to experience a biased sample of numbers of friends of others is one of a large set of related phenomena. If there is any variation in college class sizes, then students experience the average class size as being larger than it is. They experience a higher average class size than exists for the college because many students experience the large classes, while few students experience the small classes.

	Friendship Paradox		
	Number of Friends x_i	Total number of Friends of His Friends Σx_j	Mean Number of Friends of His Friends $(\Sigma x_j)/x_i$
John	3	8	8/3
Joe	4	8	2
Mark	1	4	4
Bob	2	7	3.5
Frank	2	7	3.5
Total	12	34	15.67
Average	12/5=2.4	34/12= 2.83	15.67/5=3.13

He further stated: "Whether paradoxical or not, it is important to recognize that the experiences of class sizes have a reality of their own. The fact that many individuals experience larger average class sizes (large college classes, crowded expressways, populous cities, large families, etc.) may be more sociologically and practically significant than the objective average." The recognition of the different ways that people experience the same objective situation can help us understand conflicts of interest among different individuals.

According to Science Daily (Sept. 2010), Nicholas Christakis, professor of medicine, medical sociology, and sociology at Harvard University, and James Fowler, professor of medical genetics and political science at the University of California, San Diego, used the paradox to study the 2009 flu epidemic among 744 students. As the 2009 influenza season approached, they constructed a social network that contacted 319 Harvard undergraduates, who in turn named a total of 425 friends. Monitoring the two groups through self-reporting and data from Harvard University Health Services, the researchers found that, on average, the friends group manifested the flu roughly two weeks in advance of the randomly chosen group, and a full 46 days prior to the epidemic peak in the whole population. Such finds are statistically significant (Christakis and Fowler, 2010).

Christakis and Fowler explained: Just as they come across gossip, trends, and good ideas sooner, the people at the center of a social network are exposed to diseases earlier than those at the margins. Traditionally, public health officials often track epidemics by following random samples of people or monitoring people after they get sick. The Friendship Paradox has suggested an effective way to find out which parts of the country are going to get the flu first and provides a way we can get ahead of an epidemic of flu, or potentially anything else that spreads in networks. Of course, a model is required to generalize results from a small network of friends to the whole population.

1.2.5   *Paradox of Pesticides*

Pesticides are a paradox. They are essential to the agricultural economy, helping make possible the abundant, economical food supply mankind values. Yet they are also inherently toxic. Their purpose typically to kill pests. This leads to another interesting paradox. That is, to apply pesticide to a pest, one may in fact increase its abundance. This happens when the pesticide upsets the natural predator-prey dynamics in the ecosystem. The paradox can only occur when the target pest has a naturally occurring predator that is equally affected by the pesticide. The paradox is the mathematical property of the so-called Lotka-Volterra equation, which models the rate of change of each respective population as a function of the other organism's population and pesticides.

The paradox has been documented repeatedly throughout the history of pest management. Predatory mites, for example, naturally prey upon phytophagous mites, which are common pests in apple orchards. Spraying the orchards kills both mites, but the effect of diminished predation is larger than the pesticide's, and phytophagous mites increase in abundance (Lester, Thistlewood, and Harmsen, 1998). The paradox has also been seen on rice, as documented by the International Rice Research Institute, which noted significant declines in pest populations when farmers stopped applying pesticide (Sackville Hamilton, 2008).

The Paradox of the Pesticides implies the need for more specialized pesticides that are tailored to the target pest. If the pesticide can effectively reduce only the prey population, the predator population will remain largely unaffected except for the change in its food supply. Broad-spectrum pesticides are more likely to exemplify the paradox and cause an increase in target pest population by killing predators as well. In certain cases, however, where the predator is closely related to the target pest, even narrow-spectrum pesticides may be insufficient (wikipedia.org).

1.2.6 *Hat Puzzle and Coding Theory*

The hat puzzle can be viewed as a strategy question about a cooperative game. The problem is also connected to algebraic coding theory. The hat puzzle can stated as follows:

A group of $N(> 2)$ players are wearing hats. The hats are colored black or white, and there is at least one hat of each color. Each person can see the color of every other player's hat, but not that of their own. Before the game, the players can communicate their common strategy. No communication between the players is allowed after the hats are donned, except for a single verbal guess as to the color of one's own hat. How accurate can the guesses of the players be regarding the color of their own hat?

Surprisingly, every player can guess correctly if all follow this strategy (wikipedia.org):

(1) Count the numbers b of black hats and w of white hats that you see.
(2) Wait $c = \min(b, w)$ seconds. Then:
(3) If nobody has yet spoken, guess that your hat is black if you can see fewer black hats than white hats, or white if you can see fewer white hats than black hats.
(4) If some has already spoken, guess that your hat is of the opposite color to that of the first person to speak.

Let's explain why:

Suppose that in total there are B black hats and W white hats. There are three cases.

Based on the strategy above, if $B = W$, then those players wearing black hats see $B - 1$ black hats and B white hats, so wait $B - 1$ seconds and then correctly guess that they are wearing a black hat. Similarly, those players wearing a white hat will wait $W - 1$ seconds before guessing correctly that they are wearing a white hat. So all players make a correct guess at the same time.

If $B < W$, then those wearing a black hat will see $B - 1$ black hats and W white hats, whilst those wearing a white hat will see B black hats and $W - 1$ white hats. Since $B - 1 < B \leq W - 1$, those players wearing a black hat will be the first to speak, guessing correctly that their hat is black. The other players then guess correctly that their hat is white.

The case where $W < B$ is similar.

Hat puzzles of various forms have been contemplated in coding theory, discrepancy problems, and autoreducibility of random sequences (Ebert, Merkle, and Vollmer, 2003; Feige, 2004). Immorlica (2005) exhibits a connection between hat puzzles and truthful mechanism design. He draws an analogy between agents' bids and players' hat colors, and then uses solutions to the hat puzzle to compute an offering price for each agent by setting his price equal to the corresponding player's guess. As a player must guess his hat color without observing his own hat, the mechanism we design will be bid-independent and hence truthful.

1.2.7   *Inspection Paradox and Machine Repair*

The Inspection Paradox is also called the Waiting Time Paradox. It goes like this (Leviatan, 2002; Ross, 2003, p. 437—440): "If busses run exactly every 20 minutes, then a traveler arriving at a bus stop at a random time will have to wait 10 minutes on average. The average amount of time that passed from the previous bus is also 10 minutes. Now consider a lonely road where cars pass according to a Poisson process (a reasonable assumption) at a rate of 3 per hour. A hitchhiker arrives there at a random time. By assumption, the average time between two consecutive cars is 20 minutes. The alarming news is that our hitchhiker will have to wait 20 minutes on average for the next car to pass. Moreover, the average time from the previous car is also 20 minutes!" What happened to the additive property of expectation? The solution lies in the fact that, according to the above procedure, longer periods have larger probabilities of being selected.

The phenomenon can be formally modeled using a renewal process. A renewal process is a generalization of the Poisson process. In essence, the Poisson process is a continuous-time Markov process on the positive integers, which has independent identically distributed holding times at each integer i before advancing to the next integer, $i + 1$. Counterintuitively, if we wait some predetermined time t and then observe how large the renewal interval containing t is, we should expect it to be typically larger than a renewal interval of average size.

Renewal processes can be used to solve many practical problems. Suppose Mike, the entrepreneur, has n machines, whose operational lifetime are uniformly distributed between zero and two years. Mike may let each machine run until it fails with replacement cost $2600; alternatively, he may replace a machine at any time while it is still functional at a cost of $200. What is his optimal replacement policy? The solution can be found using a renewal-reward processes, and it is to replace each machine every eight months, unless it fails first (wikipedia.org). Renewal processes can also be used to solve the insurance ruin problem and determine the probability distribution of ruin time (Ross 2003).

The Inspection Paradox has also been used in genomics (whole-genome analysis). One of the major challenges of modern biology is distinguishing meaningful patterns from the random fluctuations of DNA sequences resulting from chromosome shuffling in each generation. Due to the Inspection Paradox, a disease-causing mutation is more likely to be found in a large recombination interval. As a result, causal genetic variants are more likely to be found in larger intervals as a consequence of sampling bias. According to the paradox, the interval containing a fixed point (the causal gene variant) is about double the length of an interval not subject to this constraint. But this average doubling of length is attenuated or neutralized at the ends of chromosomes, where the distribution of interval sizes gradually returns to normal. Jeanpierre (2008) studied the consequences of sampling biases for haplotype patterns in large studies of many families and of an individual family.

1.2.8 *Paradox in Axiom System*

Mathematics as a field of endeavor has increasingly distanced itself from its empirical roots and has become an axiomatic science. This development path of mathematics is intertwined with, and paralleled by, a change in the nature and role of axioms and axiomatization. Originally understood as the articulation of self-evident, generally agreed-upon propositions pertaining to external objects, phenomena, or principles being studied, axiomatization has evolved into the activity of building a field of knowledge as a purely deductive system with propositions. Thus, for example, the axiom system of Euclidean geometry has evolved from a set of presumably self-evident certainties concerning a realm of external objects such as points, straight lines, and angles, into a set of propositions intended as complete definitions of the concepts of "point," "straight line,' "angle," etc., and aimed at recasting the study of geometric objects as a purely deductive pursuit (math.mind-draft.com).

One of the most significant works that changed the face of mathematics forever is Kurt Gödel's incompleteness theorem (Gödel, 1931), in which it is proved that any formalized system of mathematics always contains statements that are undecidable or provable. Thus, there are certain inherent limitations in any axiom system.

Gödel's theorems were surprising to many. They differed from many other famous mathematical results in the way that they came completely unexpectedly and did not require particularly difficult mathematical thought to understand, nor much knowledge beyond that of basic arithmetic and logic. Though complex and insightful, they do not overwhelm the reader with advanced mathematical concepts, and are demonstrated by surprisingly concise proofs (see Chapter 4). For this reason, Gödel's work stands out from many other groundbreaking mathematical proofs.

Gödel's incompleteness theorem exemplifies the utility of the self-reference paradox, i.e., the Liar's Paradox. The paradox says: "I am lying." If I am, then I am lying and at the same time telling the truth that I am lying. So the statement is self-contradictory, a paradox. What Gödel did essentially is to construct a Liar's Paradox in arithmetic: "I am unprovable." The Gödel's incompleteness theorem can also be proved via Berry's Paradox.

Gödel's incompleteness theorem is closely related to computability theory in computer science. Stephen Cole Kleene (1943) presented a proof of Gödel's incompleteness theorem using basic results of computability theory. One such result shows that the halting problem is unsolvable: There is no computer program that can correctly determine, given a program P as input, whether P eventually halts when run with some given input. Kleene showed that the existence of a completely effective theory of arithmetic with certain consistency properties would force the halting problem to be decidable, a contradiction.

1.2.9 *Paradox and Electric Bell*

One of the proudest accomplishments of my childhood was creating an electric bell, though later I found that it was just a reinvention. Other reinventions I remember are discovering some of the interesting properties of the number 9 and the solution for a general quadratic equation. However, I am still proud of myself even knowing they are just reinventions because I did them all in the "dark age" of China: the 10 years of the Chinese Cultural Revolution.

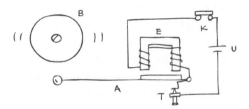

THE PRINCIPLE OF AN ELECTRIC BELL WITH DIRECT CURRENT

An electric bell is a mechanical bell that functions by means of an electromagnet. When an electric current is applied, it produces a repetitive buzzing or clanging sound. Electric bells are still widely used in telephones, fire alarming, school bells, and doorbells, but they are now being widely replaced with electronic sounders.

The principle that makes a bell work is essentially a paradox: The first action causes a second action that cancels the first action; the cancellation of the first action cancels the second action, the cancellation of the second triggers the first again,..., the chain of actions continuing forever. The two physical actions take a limited time, resulting in alternations repeatedly. Specifically, the bell (B) can produce sound when it is struck by a spring-loaded arm with a metal ball on its end called a clapper (A), actuated by an electromagnet (E). In its rest position, the clapper is held a short distance away from the bell by its springy arm. When an electric current is passed through the winding of the electromagnet it creates a magnetic field that attracts the iron arm of the clapper, pulling it over to give the bell a tap. This opens a pair of electrical contacts (T) attached to the clapper arm, cutting off the electric circuit. The magnetic field of the electromagnet collapses, and the clapper springs away from the bell. This closes the contacts again, allowing the current to flow to the electromagnet, so that the magnet pulls the clapper over to strike the bell again. This cycle repeats rapidly, resulting in a continuous ringing.

Any negative feedback system can thus be viewed as a paradox. For example, the speed stabilizer of the classic (mechanical) record player is also based on a basic paradox: An increase in rotation speed will lead to an increase in the moment of inertia that will reduce the speed. Just as when a ballet dancer opens her arms, the moment of inertia will increase and slow down her spin.

1.2.10 *Paradox and Message Encryption*

The commonly used encryption method is based on the Paradox of Cryptography, a countintuitive fact that, while finding tremendously large prime numbers is computationally easy, factoring the product of two such numbers is at present computationally infeasible.

Online exchanges of sensitive data over networks need to be encrypted to prevent the data from being disclosed to or modified by unauthorized parties. The encryption must be done in such a way that decryption is only possible with knowledge of a secret decryption key, which is only known by authorized parties.

A password could be used to derive an encryption key, but how do you protect the password from interception when you're first setting it up? In

1976, Whitfield Diffie and Martin Hellman introduce the concept of public-key cryptography in their landmark paper "New Directions in Cryptography." In public-key cryptography, a pair of different keys are used for encryption and decryption. The decryption key would still have to be kept secret, but the encryption key could be made public without compromising the security of the decryption key. To enable computers to encrypt data for a site, the site simply needs to publish its encryption key, for instance, in a directory. Every computer can use that encryption key to protect data sent to the site. But only the site has the corresponding decryption key, so only it can decrypt the data.

The basic idea of one-way encipherment is very simple: Two numbers can easily be multiplied by each other, e.g., the product of 101 and 211 can be calculated quickly, it is 21311; but if we want to find two integers greater than one whose product is 21311, then it will take much more time to find that 101 and 211 is the only possible solution. Naturally there are computer algorithms for factoring numbers, but in the case of a 40—50 digit number, the running time required would be millions of years. On the basis of prime number theory, a simple trapdoor function was found: The enciphering key depends only on the product of two prime numbers, whereas to decipher the ciphertext the two prime numbers have to be known also.

Let p and q be large random prime numbers. The product $n = pq$ and another random number E are the public key, (E, n). To apply the key, a sender first converts his message into a string of numbers, which he then breaks into blocks $B_1, B_2, ...$, satisfying $0 < B_i < n-1$. Let $C_i = B_i^E$ modulo n. As mentioned earlier, factoring the huge product number n is at present computationally infeasible. So, knowing only (E, n) and C_i, it is hopelessly difficult to find B_i (thus the sender's message, too). To decipher a ciphertext $C_1, C_2,...$, the user needs to employ n and a secret deciphering key D, where D is the multiplicative inverse of E modulo $(p-1)(q-1)$, that is, ED modulo $(p-1)(q-1)$ is equal to 1. (The product $(p-1)(q-1)$ is the number of integers between 1 and n that have no common factor with n; $1 < E < (p-1)(q-1)$ and no common factor between E and $(p-1)$ or $(q-1)$.)

After all this the receiver can easily obtain the B_i:

$$C_i^D = \left(B_i^E\right)^D = B_i \bmod n.$$

This method was designed by Rivest, Shamir, and Adleman and is called the RSA system. The RSA system has also been used in Digital Signature. A digital signature is used to protect the data it sends from modification.

1.3   Mathematical Paradox

1.3.1   *Achilles and The Tortoise*

Zeno's paradoxes are a set of problems generally thought to have been devised by Zeno of Elea to support Parmenides's doctrine that "all is one" and that, contrary to the evidence of our senses, the belief in plurality and change is mistaken, and in particular that motion is nothing but an illusion. Zeno took on the project of creating these paradoxes because other philosophers had created paradoxes against Parmenides's view (Sainsbury, 1995). One of the most famous paradoxes is Achilles and The Tortoise. It goes like this:

ACHILLES & THE TORTOISE RACE

In a footrace with a tortoise, Achilles allows his opponent a head start of 100 meters. We can suppose they are running at different but constant speeds. After some finite time, Achilles will have run 100 meters, bringing him to the tortoise's starting point. During this time the tortoise has run a much shorter distance, say 10 meters. It will then take Achilles some further time to run that distance, by which time the tortoise will have advanced farther; and then more time still to reach this third point, while the tortoise moves ahead. Thus, whenever Achilles arrives somewhere the tortoise has already been, he still has farther to go. This will go on and on. No matter how much faster Achilles runs he must first reach where the tortoise has already been; thus he can never overtake the tortoise. In this argument, it is implicitly assumed that the sum of an infinite series of numbers will always be infinitely large, which now is well known to be untrue. Imagine that one cuts a foot-long ruler into two parts with equal lengths, then takes one of them and further cuts it into two equal pieces; this process continues so that we have an infinite number of pieces with lengths $\frac{1}{2}, \frac{1}{2^2}, ..., \frac{1}{2^n}, ...$ However, the sum of all the lengths is 1 foot, the original length of the ruler.

1.3.2 *Hilbert's Paradox of Grand Hotel*

Throughout the whole history of humanity many brilliant thinkers studied problems related to the idea of infinity. Hilbert's Paradox of the Grand Hotel is one of them. The paradox is a mathematical paradox about infinite sets presented by German mathematician David Hilbert. It is counterintuitive because it implies that a part can be equal to the whole.

Suppose the hotel is already full of guests but a new guest arrives and wishes to be accommodated in the hotel. Because the hotel has infinitely many rooms, we can move the guest occupying room 1 to room 2, the guest occupying room 2 to room 3, and so on, and fit the newcomer in room 1. By repeating this procedure, it is possible to make room for any finite number of new guests.

It is also possible to accommodate a countably infinite number of new guests: Just move the person occupying room 1 to room 2, the guest occupying room 2 to room 4, and in general, room n to room $2n$, and all the odd-numbered rooms will be free for the new guests.

For any countably infinite set, there exists a bijective function that maps the countably infinite set to the set of natural numbers, even if the countably infinite set contains the natural numbers. For example, the set of rational numbers—those numbers which can be written as a ratio of integers—contains the natural numbers as a subset, but is no bigger than the set of natural numbers since the rationals are countable: There is a bijection from the naturals to the rationals.

1.3.3 *Berry's Paradox of the Smallest Integer*

Berry's Paradox is also a self-referential paradox. There are several versions of the paradox, the original version of which was published by Bertrand Russell and attributed to Oxford University librarian Mr. G. Berry. In the form stated by Russell (1908), the paradox notes that the least integer not nameable in fewer than nineteen syllables is itself a name consisting of eighteen syllables; hence the least integer not nameable in fewer than nineteen syllables can be named in eighteen syllables, which is a contradiction."

The version from wikipedia.org discusses "the smallest positive integer not definable in under 11 words." Since there are finitely many words, there are finitely many positive integers that are defined by phrases of under 11 words.

Since there are infinitely many positive integers, this means that there are positive integers that cannot be defined by phrases of under eleven words. By the well-ordering principle, if there are positive integers that satisfy a given property, then there is a smallest positive integer that satisfies that property. Therefore, we can let N be the smallest positive integer satisfying the property of being not definable in under 11 words. In other words, N is defined by the expression "the smallest positive integer not definable in under 11 words." The above expression defining N is only 10 words long, so N is definable in under 11 words.

The contradiction arises because there are infinitely many positive integers but there are only a finite number of combinations of 11 or fewer English words; it is impossible to make one-to-one mapping between the two sets. If all other combinations of words uniquely map to integers, then we actually use the same expression, "the smallest positive integer not definable in under 11 words" to define (infinitely) many different positive integers.

George Boolos (1989) built on a formalized version of Berry's Paradox to prove Gödel's incompleteness theorem in a new and much simpler way than Gödel did.

1.3.4   *Torricelli's Trumpet of Finiteness and Infinity*

For a two-dimensional object, if both dimensions are bounded, then the area of the object is bounded. If one of the dimensions is unbounded, we may image intuitively that the area will be unbounded too. However, this is not necessarily true. For instance, the area under the curve $f(x) = 1/x^2$ within the range of $x > 1$ is 1. Can you now imagine a single three-dimensional object that encompasses both a finite volume but an infinite surface area? Torricelli's Trumpet is a classic example of such an object.

In his brief life (1608—1647), the interests of Italian mathematician and physicist Evangelista Torricelli ranged from pure mathematics to experimental physics. Castelli, who was a former student of Galileo, thought so highly of Torricelli's genius that he made him his secretary, a post the younger man held for 6 years. In mathematics, he took delight in problems involving the use of infinitesimal methods.

Torricelli's trumpet, a figure invented by Evangelista Torricelli in 1964, has an infinite surface area, but a finite volume. The trumpet is formed by taking the area under $y = 1/x$, and rotating it in three dimensions about the x-axis. The discovery was made using Cavalieri's principle before the invention of calculus, but today calculus can be used to calculate the volume (V) and surface area (A) of the trumpet beyond $x = 1$. It turns out that

$V = \pi$ and $A = \infty$.

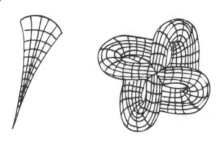

TORRICELLI'S TRUMPET
& MORIN SURFACE (SMALE'S PARADOX)

Torricelli's proof is very different and more complex. At the time, this result generated what appeared to be a paradox, namely: It seems it would take an infinite amount of paint to coat the outside surface of Gabriel's horn, but as its volume is finite, the interior surface could be coated with a finite amount of paint. The paradox is resolved by considering its assumptions and noting that much of the interior of the horn would be inaccessible, because paint molecules have thickness, and even if they did not, it would take an infinite amount of time for the paint to run down the infinite length of the horn (www.robertnowlan.com).

If this paradox does not surprise you enough, the next one will. In differential topology, Smale's Paradox states that it is possible to turn a sphere inside out in a three-dimensional space with possible self-intersections but without creating any crease, a process often called sphere eversion (eversion means "to turn inside out"). For visualizations of eversion, you can search on google.com or youtube.com.

1.3.5 *Barber's Paradox in Naive Set Theory*

The Barber Paradox (Russell's Paradox), discovered by Bertrand Russel in 1901, shows that an apparently plausible scenario is logically impossible. Suppose there is a town with just one male barber who claims that he will shave every man in the town who never shaves himself, but will not shave any man who shaves himself. This sounds very logical until the following question is asked: Does the barber shave himself?

If the barber does not shave himself, he must abide by his rule and shave himself. However, if he does shave himself, then according to the rule he will not shave himself.

Russell's Paradox, restated in the context of so-called "naïve" set theory, exposed a huge problem and changed the entire direction of twentieth century mathematics.

THE BARBER'S PARADOX

There are two types of sets in naïve set theory. For the first type of sets, the set is not a member of itself. For instance, the set of all nations is not a nation; the set of all men is not a man. For the second type of set, the set is a member of itself. For instance, the set of everything which is not men is a member of itself. Now, what about the set of all sets that are not members of themselves? Is it a member of itself or not? If it is, then it isn't, and if it isn't, then it is ... Just like the barber who shaves himself, but mustn't, and therefore doesn't, and so must!

Russell's Paradox means that there is a contradiction at the heart of naïve set theory. That is, there is a statement S such that both itself and its negation (not S) are true, where the statement S is "the set of all sets that are not members of themselves contains itself." The problem is that once there is such a contradiction, one can prove anything being true using the rules of logical deduction! This is how it goes:

Suppose if S is true, then Q is true, where Q is any other statement. But "not S" is also true, so Q is true no matter what it is. The paradox raises the frightening prospect that the whole of mathematics is based on shaky foundations, and that no proof can be trusted. The Zermelo-Fraenkel set theory with the axiom of choice was developed to escape from Russell's Paradox.

1.3.6   *Fermat's Last Theorem*

Fermat's last theorem in number theory states that no three positive integers a, b, and c can satisfy the equation $a^n + b^n = c^n$ for any integer value of n greater than two. This theorem was first conjectured by Pierre de Fermat in 1637, but he left no proof of the conjecture except for the case of $n = 4$. The unsolved problem stimulated the development of algebraic number theory in the 19th century and the proof of the modularity theorem in the 20th.

The list of twenty-three problems in mathematics published by German mathematician David Hilbert in 1900 has been greatly inspiring generations of mathematicians. He is one of the greatest mathematics of all time. Hilbert was asked to submit the proof of Fermat's last theorem himself to win the Wolfskehl Prince, but he laughed it off by saying, "Why should I kill the goose that lays the golden egg?" The "golden egg" he had in mind was the interest on the principal, amounting to 5000 marks per annum (Mehra and Rechenberg, 2000).

A concernstone for the proof of the conjecture came in 1984 from Gerhard Frey. He suggested the approach of proving the conjecture through the modularity conjecture for elliptic curves. Building on work of Ken Ribet, Andrew Wiles succeeded in proving enough of the modularity conjecture to prove Fermat's last theorem. Frey's proof of Fermat's last theorem required two steps. First, it was necessary to show that Frey's intuition was correct, that the above elliptic curve is always nonmodular. Frey did not succeed in proving this rigorously; the missing piece was identified by Jean-Pierre Serre. This missing piece, the so-called "epsilon conjecture," was proved by Ken Ribet in 1986. Second, it was necessary to prove a special case of the Taniyama— Shimura conjecture, raising it from a mere conjecture to a theorem. This special case (for semistable elliptic curves) was proved by Andrew Wiles in 1995. Summarizing, the epsilon conjecture showed that any solution to Fermat's equation could be used to generate a nonmodular semistable elliptic curve, whereas Wiles' proof showed that all such elliptic curves must be modular. This contradiction implies that there can be no solutions to Fermat's equation, thus proving Fermat's last theorem.

1.3.7   *Borel's Normal Number*

Borel's normal numbers are numbers of a very strange type. Each of them consists of a sequence of digits and is defined by the probability distribution of digits. A number is normal in base b if every sequence of k symbols in the letters $0, 1, ..., b - 1$, occurs in the base-b expansion of the given number with the expected frequency b^{-k}. For $k = 1$, a normal number with base-b expansion has each digit appearing with average frequency tending to $1/b$. We can think of numbers normal in base 2 as those produced by flipping a fair coin, recording 1 for heads and 0 for tails.

A normal number is an irrational number for which any finite pattern of numbers occurs with the expected limiting frequency in the expansion in a given base (or all bases). For example, for a normal decimal number, each digit 0—9 would be expected to occur $1/10$ of the time, each pair of digits 00—99 would be expected to occur $1/100$ of the time, etc. A number that is normal in base-b is said to be b-normal. A number that is b-normal for every $b = 2, 3, \ldots$ is said to be absolutely normal.

Despite the abundance of such numbers, it is exceedingly difficult to find specific examples. It is known that the Champernowne number $0.123456789101112131415 \cdots$ is normal in base 10. It is widely believed that the numbers $\sqrt{2}$, π, and e are normal, but a proof remains elusive (Marques, 2008).

A given infinite sequence is either normal or not normal, whereas a real number, having a different base-b expansion for each integer $b \geq 2$, may be normal in one base but not in another. All normal numbers in base r are normal in base s if and only if $\log r / \log s$ is a rational number (wikipedia.org).

1.3.8 *Paradoxes of Enrichment*

Regime shifts in complex systems are characterized by abrupt transitions between alternative persistent states. Decreasing resilience (the ability of a system to resist change) has been suggested as a useful leading indicator of regime shifts in such systems. The "Paradox of Enrichment" is a classic ecological model predicting that ecosystem enrichment can lead to a regime shift to an undesirable state (Chisholm and Filotas, 2009). The Paradox of Enrichment is a term from population ecology coined by Michael Rosenzweig in 1971. As the carrying capacity increases, the equilibrium of the dynamic system becomes unstable. The Paradox of Enrichment is a phenomenon called bifurcation, which can be obtained by the modified Lotka—Volterra equation.

There are several equilibria. The first equilibrium corresponds to the extinction of both predator and prey, the second one to the extinction of the predator, and the third to coexistence. Beyond a certain critical point, the system undergoes a so-called Hopf bifurcation; i.e., increasing the carrying capacity of the ecological system beyond a certain value leads to dynamic instability and extinction of the predator species. However, so far there is no empirical evidence to support the Paradox of Enrichment. It is only a theoretical result that heavily depends on the mathematical model.

1.3.9 *Logistic Model and Chaos*

The logistic model is widely used and often cited as an archetypal example of chaotic behavior that can arise from very simple nonlinear equations. If x_n is the ratio of the existing population at year n to the maximum possible population (the carrying capacity), the same quantity next year can be modeled using the logistic equation

$$x_{n+1} = rx_n\left(1 - x_n\right),$$

where r is a positive number, representing a combined rate for reproduction and starvation.

The logistic equation cannot be solved exactly in general. However, for $r = 2$ and 4, exact solutions can be found. The solution when $r = 2$ is

$$x_n = \frac{1}{2} - \frac{1}{2}\left(1 - 2x_0\right)^{2^n}, \quad x_0 \in [0, 1).$$

We can see that as n goes to infinity, x_n goes to the stable fixed point $1/2$, unless $x_0 = 0$.

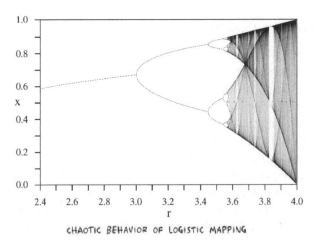

CHAOTIC BEHAVIOR OF LOGISTIC MAPPING

The solution when $r = 4$ is $x_{n+1} = sin^2(2^n\theta\pi)$. For rational θ, after a finite number of iterations, x_n maps into a periodic sequence. But for an irrational θ, x_n is nonperiodic—it never repeats itself. This solution demonstrates the two key features of chaos—stretching and folding: The factor 2^n shows the exponential growth of stretching, which results in sensitive dependence on initial conditions, while the squared sine function keeps x_n folded within the range $[0, 1]$.

The behavior of the logistic model varies with the value of the parameter r, as outlined below (Pickover, 2006, p. 54—61; Mitchell, 2009, p. 36—39; wikipedia.org):

- When $0 \leq r < 1$, the population will eventually die out, independent of the initial population.
- When $1 \leq r < 2$, the population will quickly approach the value 1, independent of the initial population.
- When $2 \leq r < 3$, the population will also eventually approach the same value.
- When $3 \leq r < 1 + \sqrt{6}$, from almost all initial conditions the population will approach permanent oscillation between two values. These two values are dependent on r.
- When $1 + \sqrt{6} \leq r < 3.57$ (approximately), from almost all initial conditions the population will approach permanent oscillation among 4, 8, 16, ... values.
- At r approximately 3.57 chaos commences, at the end of the period-doubling cascade.

The logistic function is a simple model for learning the concepts of chaos and bifurcation. The bifurcation diagram is a fractal with self-similarity: If you zoom in on any branch, the situation nearby looks like a shrunken and slightly distorted version of the whole diagram.

1.3.10 *Paradoxes Related to Fixed Points*

In mathematics, a fixed point (fixed-point or an invariant point) of a function is a point that is mapped to itself by the function. In other words, we call x a fixed point if $f(x) = x$. In general, if for an integer n, $f^n(x) = f^{n-1}(f(x)) = x$, we say x is a periodic point of period n. A surprising theorem about periodic points is that if a continuous function has a point of period 3 it must have points of all periods 1,2, ... There are many counterintuitive examples of fixed points. Here we list just a few (mindyourdecision.com):

One morning at 6 am, a monk began climbing a tall mountain; he reached the top at 8 pm. The next morning at 6 am, he descended the mountain along the same path and reached the bottom at 8 pm. There is some spot on the path that the monk occupied at precisely the same time of day for both trips, no matter what he did during the trips going up or down.

A MONK CLIMBS A MOUNTAIN

You are playing with a can of soda. However, no matter how well you shake it up, at least one point in the can was not moved (relative to the can): It ended up in the same spot as where it started. This result is due to Brouwer's fixed point theorem. Similarly, no matter what the weather patterns are, there is at least one spot on earth where the wind is not blowing. The point where there is no wind blowing is a fixed point. This result is, essentially, the hairy ball theorem.

In many fields equilibria and stability are fundamental concepts that can be described in terms of fixed points. For example, Nobel prize winner in Economics John Nash proved that any zero-sum rational game has a solution (Nash equilibrium) at which no party in the game will make a move unilaterally. Nash relied on the Kakutani fixed point theorem in his proof. Nash's existence proof was not appreciated at the time of publication in 1950. The polymath John von Neumann is believed to have quipped, "That's trivial, you know. That's just a fixed point theorem." In physics (the theory of phase transitions), linearization near an unstable fixed point led to Wilson's Nobel prize-winning work inventing the renormalization group, and to the mathematical explanation of the term "critical phenomenon."

In compilers of computer languages, fixed point computations are used for whole program analyses, which are often required to do code optimizations. The vector of PageRank values of all webpages is the fixed point of a linear transformation derived from the World Wide Web's link structure. The concept of fixed point can also be used to define the convergence of a function.

1.3.11   *Paradox of π: Möbius Band*

The Möbius strip (or band) is familiar to everyone. Its popularity and its applications can be found in mathematics, magic, science, art, music, engineering, and literature. The ubiquitous symbol for recycling is one form of the Möbius band (Pickover, 2006). The strip is a surface with only one side. A model can easily be created by taking a paper strip and giving it a half-twist, and then joining the ends of the strip together to form a loop.

Möbius (1790—1868) was the epitome of the absent-minded professor. He was shy and unsociable. His most impressive discovery, the Möbius band, was made when he was almost 70, and all the works found among his papers after his death show the same excellence of form and profundity of thought (Pickover, 2006). Gardner (Martin Gardner, 1976 from Pickover, 2006) said it perfectly: "Even a great mathematician is almost always unknown to the public. His "adventures" are usually so confined to the interior of his skull that only another mathematician cares to read about them."

The Möbius strip has numerous fascinating properties: Cutting it along the center line yields one long strip with two full twists in it, rather than two separate strips. A strip with an odd number of half-twists, such as the Möbius strip, will have only one surface and one boundary. A strip twisted an even number of times will have two surfaces and two boundaries.

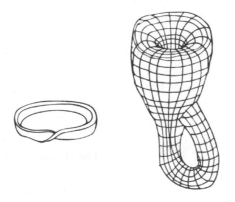

MOBIUS BAND & THE KLEIN BOTTLE

If a blind ant crawls along the middle of a Möbius band, it will eventually return to its starting point and the path it takes will be double the length of the original strip. Suppose the ant didn't know it was traveling on a circular Möbius strip. It just remembered crawling a constant distance from some central point, which it thinks is the center of a circle. It uses the relationship between the circumference C and the diameter d:

$$C = \pi d,$$

but the π is 6.28... double the value of the familiar 3.14... The question is, how do we know we are not living in a "3D Möbius strip"?

More interestingly, if we glue two parallel Möbius strips together along their edges, it will become a "strange" geometrical object: the Klein bottle. This cannot be done in ordinary three-dimensional Euclidean space without creating self-intersections (Gardner, 2009; Pickover, 2006).

1.3.12   *Dimensional Paradox and Network Robustness*

Fractals are graphs with self-similarity, at least stochastically. Such a similarity is usually defined by a simple and recursive algorithm. The notion of fractals started in 1872 when Karl Weierstrass gave an example of a function with the non-intuitive property of being everywhere continuous but nowhere differentiable. In 1904, Helge von Koch gave a more geometric definition of a similar function, which is now called the Koch curve (Pickover, 2009, p. 310). The later concept of statistical self-similarity and fractional dimension are built on earlier work by Lewis Fry Richardson. In 1975, Mandelbrot coined the word *fractal* to denote an object whose (Hausdorff) fractal dimension is greater than its topological dimension. He illustrated his mathematical definition with striking computer-constructed visualizations (wikipedia.org).

A binary tree is an example of a fractal object. In the biological world, fractal structures such as venular and arterial trees cannot be characterized by geometric self-similarity; rather, they possess *statistical self-similarity*. The fractal is statistically self-similar since the characteristics (such as the average value or the variance, or higher moments) of the statistical distribution for each small piece are proportional to the characteristics that concern the whole object (Macheras and Iliadis, 2006). Fractal study can be helpful in many areas. For instance, Macheras and Iliadis (2006, p. 31—37) study the fractal structures of the cardiovascular system and their applications in pharmacokinetics in new drug development.

Scaling laws for fractal objects state that if one measures the value of a geometric characteristic $\theta(w)$ on the entire object at resolution w, the corresponding value measured on a piece of the object at finer resolution $\theta(rw)$ with $r < 1$ will be proportional to $\theta(w)$:

$$\theta(rw) = k\theta(\omega)$$

where k is a proportionality constant that may depend on r.

The above-delineated dependence of the values of the measurements on the resolution applied suggests that there is no single true value of a measured characteristic. A function $\theta(\omega)$ satisfying the scaling law is the power law function: $\theta(\omega) = \beta\omega^{\alpha}$.

JULIA SET

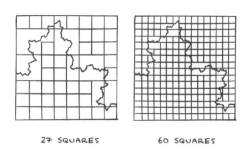

27 SQUARES   60 SQUARES

FRACTAL DIMENSION = LOG(60/27)/LOG(2) = 1.152...

The fractal dimension D is defined as $\ln(k)/\ln(r)$ as r approaches zero. Suppose M $\omega \times \omega$ squares are required to cover the fractal or N smaller $\frac{\omega}{2} \times \frac{\omega}{2}$ squares are required to cover the same fractal; then the (Hausdorff) fractal dimension D is the limit of $\ln(N/M)/\ln(2)$ as ω approaches 0 (Appignanesi, 2006).

Fractals are closely related to scale-free networks by the power law. Many people believe that World Wide Web links, biological networks, and social networks are scale-free networks. A scale-free network is a network whose degree distribution follows a power law, at least asymptotically. That is, the fraction $P(k)$ of nodes in the network having k connections to other nodes is proportional to k^r, where the parameter r typically ranges from 2 to 3.

The scale-free property strongly correlates with the network's robustness to failure. It turns out that the major hubs (the highest-degree nodes) are closely followed by smaller ones. These, in turn, are followed by other nodes with an even smaller degree, and so on. This hierarchy allows for a fault-tolerant behavior. If failures occur at random and the vast majority of nodes are those with small degree, the likelihood that a hub would be affected is almost negligible. Even if a hub-failure occurs, the network will generally not lose its connectedness, due to the remaining hubs. On the other hand, if we choose a few major hubs and take them out of the network, the network is turned into a set of rather isolated graphs. Thus, hubs are both a strength and a weakness of scale-free networks. These properties have been studied analytically using percolation theory by Cohen et al.(2000, 2001) and by Callaway et al. (2000) (wikipedia.org)

It is easy to see that people tend to form communities, i.e., small groups in which everyone knows everyone. One can think of such a community as a complete graph. In addition, the members of a community also have relationships with people outside that community. Some people, however, are connected to a large number of communities (e.g., celebrities, politicians).

The random removal of even a large fraction of vertices impacts the overall connectedness of a lower-degree nodes network very little, suggesting that such topologies could be useful for security, while targeted attacks destroy the connectedness very quickly.

1.3.13   *The Possibility of The Impossible*

The Incredible Square

Pickover (2001) gives his readers many interesting mathematical examples. Here is an example: The two squares below both have the amazing property that the sums for the rows, columns, and diagonals are 242. Even more surprising, each entry in the left square is the reverse of the corresponding entry in the right square (Pickover, 2001, p. 238). Can you find another square that has a similar property?

96	64	37	45
39	43	98	62
84	76	25	57
23	59	82	78

69	46	73	54
93	34	89	26
48	67	52	75
32	95	28	87

Pretzel Transformation Puzzle

If the object on the top-left of the figure is made of a very elastic material, is it possible to unchain the two loops (the figure on top-right of the figure) through a continuous transformation (Pretzel Transformation) without breaking one of the loops? To me the answer was obvious: It is impossible. However, surprisingly, the solution is incredibly simple: The arrows show the solution (Pickover, 2006, p. 204).

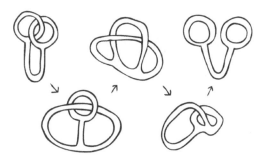

TRANSFORM THE LINKED RINGS
WITHOUT CUTTING A RING

1.4 Probabilistic and Statistical Paradoxes

There are a large variety of paradoxes in probability and statistics. Some of them are interesting mainly from a historical point of view, as the theory has already been adapted to resolve them, while others hide common misconceptions in a very subtle and tricky way. Here we just discuss some simple ones. More challenging examples will be discussed in future chapters.

1.4.1 *Roulette Paradox*

Roulette is a popular casino game devised in 18th century France. In this game, players may choose to place bets on either a single number or a range of numbers, the colors red or black, or whether the number is odd or even. To determine the winning number and color, a dealer spins a wheel in one direction, then spins a ball in the opposite direction around a tilted circular track running around the circumference of the wheel. The ball eventually loses momentum and falls on to the wheel and into one of 37 (in French/European roulette) or 38 (in American roulette) colored and numbered pockets on the wheel (Wikipedia.org). The probability of stopping at each pocket is a given number. However, if we ask (Leviatan, 2002): What is the probability that the angle will be exactly any given angle, say π? From continuity and equal possible stopping at any angle, we know that this probability must be zero. This seems to imply that the roulette will not be able to come to rest. Paradoxically, the roulette will stop at some angle θ, which originally had probability zero. Therefore, we conclude: The probability of an impossible event is zero, but the probability of an event may be zero without it being impossible.

ROULETTE WHEEL

A similar paradox (Leviatan, 2002) is about a boy who decided to save the money he received. During the first half year he got 2 bills, but at the end of this period he randomly took out 1 bill. In the next 1/4 year he got 2 more bills, but at the end of this period he randomly took out 1 bill from the 3 bills in his bank. This routine is repeated infinitely many times (each period is half the length of the previous period). What is the probability that any one of the bills he got during the year will remain in his bank after a full year? Paradoxically, the probability is 0, even though it is clear that he only spent half of his money.

1.4.2   *Birthday Paradox*

The Birthday Paradox is one of the most famous and well known problems in probability theory. It states that if there are 23 or more people in a room then the probability of any two of them sharing a birthday is greater than 50%. This comes as a surprise to many. This is because, to many people, the number 23 seems very small for there to be a 50-50 chance of a pair of people sharing a birthday (Aginer and Ziegler, 2004, 157). The probability of having the same birthday among 23 people is simply calculated as

$$p = 1 - \frac{365 \cdot 364 \cdots (365 - 23 + 1)}{365^{23}} = 0.507$$

The Birthday Paradox can be equivalently stated as: "If I guess at your birthday 23 times, there is $p > 50\%$ probability that one of my guesses is right." If I repeat this for another friend, the probability that I guess one of birthday correctly is $p > 1 - (1 - 0.5073)^2 = 0.757\,25$.

The Birthday Paradox was studied for an application in cryptology by Mike Seitlheko (2010).

1.4.3 *Elevator Paradox (Chinese Box)*

The Elevator Paradox is a paradox first noted by Gamow and Stern (1958), physicists who had offices on different floors of a multi-story building. Gamow, who had an office near the bottom of the building, noticed that the first elevator to stop at his floor was most often going down, while Stern, who had an office near the top, noticed that the first elevator to stop at his floor was most often going up.

ELEVATOR PARADOX

Assume there are n floors in the building and the person is at the kth floor. The elevator can come down to the person from any of the $(n - k)$ floors and can come up from any one of the $(k - 1)$ floors below. Therefore, the probability of the elevator going down is (Gorroochurn, 2012, p. 260)

$$\frac{n - k}{(n - k) + (k - 1)} = \frac{n - k}{n - 1}.$$

Hence, for small k (a lower floor), the probability is larger, whereas for larger k (an upper floor), the probability is smaller. Interestingly, if there is more than one elevator in a building, the conclusion may not be true (Gardner 1986, Knuth 1969).

However, such an explanation is not nearly satisfactory, or even the existence of the paradox is questionable. In fact, whether the paradox exists or not depends on the travel direction of the person waiting. Suppose there are 10 floors in the building. Gamow's office is at the 4th floor, and he only uses the elevator when he leaves for home. Then he will only see, one time, the elevator going down (when one he takes it down to the first floor) each day, and more times the elevator going up. Therefore, whether he sees more times the elevator going up or down depends not only on the floor of his office, but also the numbers of times he wants to go up or down at that floor. From further analyses, we can easily conclude that it is also dependent on the usage of the elevator by other people.

1.4.4 *Paradox of A Drunken Man*

A drunken person returns home and wants to get in. There are 10 keys on his key ring, and only one fits. He randomly pulls a key and tries to open the door. If it doesn't fit, the key returns to the ring and he randomly pulls a key (out of 10) again, and again, until the door opens. Under these circumstances, the number of trials required is 1, 2, 3, . . . Which of these values is most likely? You may be surprised to find that the most likely value is 1 (Leviation, 2002).

The following simple probability calculation explain why. Suppose there are n keys ($n = 10$ is a special case). The probability of getting the right key at the first trial is simply $p_1 = \frac{1}{n}$.

The probability of getting the right key with exactly two trials is the probability of being wrong at the first trial $\left(\frac{n-1}{n}\right)$ and right at the second trial $\left(\frac{1}{n}\right)$. Due to the independence of the trials, this probability is

$$p_1 = \left(\frac{n-1}{n}\right) \cdot \left(\frac{1}{n}\right) = \frac{n-1}{n} p_1 < p_1.$$

Similarly, the probability of getting the right key with exactly three trials is the probability of being wrong at the first and second trials, but right at the third trial. In general, the probability of getting the right key with exactly m trials is

$$p_m = \left(\frac{n-1}{n}\right)^m p_1 < p_1.$$

Therefore, we conclude $p_1 > p_2 > \cdots > p_\infty = 0$. In other words, it is most likely that he will get the right key at the very first attempt, even if he is drunk. What a surprise!

1.4.5 *Saint Petersburg Paradox*

In economics, the St. Petersburg Paradox (Bernoulli 1954) is a paradox related to probability theory and decision theory. The paradox is a classic situation where a naïve decision criterion (based on the expected value only) would recommend a course of action that no rational person would be willing to take.

In the St. Petersburg game, a player pays a fixed amount to enter the game. The game is played by flipping a fair coin until it comes up tails, and the total number of flips, n, determines the prize, which equals $\$2^n$. Thus, if the coin comes up tails the first time, the prize is $\$2^1 = \$2$, and the game ends. If the coin comes up heads the first time, it is flipped again. If it comes up tails the second time, the prize is $\$2^2 = \$4$, and the game ends. If it comes up heads the second time, it is flipped again. And so on. There are an infinite number of possible "consequences" (runs of heads followed by one tail) possible. The probability of a consequence of n flips ($P(n)$) is 1 divided by $2n$, and the "expected payoff" of each consequence is the prize times its probability.

$$E = \frac{1}{2} \cdot 1 + \frac{1}{4} \cdot 2 + \frac{1}{8} \cdot 4 + \cdots = \sum_{i=1}^{\infty} \frac{1}{2} = \infty$$

The expected win for the player of this game is infinity if the casino has an unlimited amount of money and the player has an infinite time to play. However, a large payoff comes along very rarely. According to the usual treatment of deciding when it is advantageous and therefore rational to play, one should therefore play the game at any price if offered the opportunity.

The classical resolution of this paradox involved the explicit introduction of a utility function. A gain of w dollars is more significant to the poor man than to the rich man, though both gain the same amount. A common utility model is the log utility, $\ln(w)$, representing the gambler's total wealth (wikipedia.org).

1.4.6 *Paradox of the Family Tree*

In the first half of the previous century, an interesting phenomenon was noticed, namely, the gradual extinction of several famous common and aristocratic family names (Székely, 1986). The problem is generally characterized by the so-called branching process. A branching process is a Markov process that models a population in which each individual in a generation produces some random number of individuals in the next generation. A central question in the theory of branching processes is the probability of ultimate extinction. Branching processes can also be used to model other systems with similar dynamics, e.g., the propagation of neutrons in a nuclear reactor.

BRANCHING PROCESS — FAMILY TREE

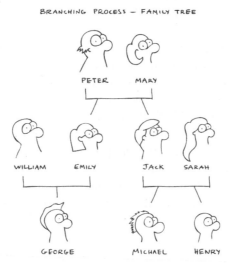

It is proved that starting with one individual in generation one, the expected size of generation n equals $\mu^{(n-1)}$, where the basic reproductive rate, μ, is the expected number of children of each individual. If μ is less than 1, then ultimate extinction has probability 1 by Markov's inequality. If μ is greater than 1, then the probability of ultimate extinction is less than 1 (Taylor and Karlin, 1998, p. 177—180).

1.4.7   *Paradox of Choice*

Freedom is the right to choose: the right to create for oneself the alternatives of choice. Without the possibility of choice, and the exercise of choice, a man is not a man but a member, an instrument, a thing. —Thomas Jefferson

The popular psychological theories affirm the positive affective and motivational consequences of having personal choice. Studies rooted in conventional wisdom have linked the pervasiveness of choice to increased levels of intrinsic motivation, greater persistence, better performance, and higher satisfaction, or simply put: the more choice, the better. However, observed in many cases is the paradox that more choices may lead to a poorer decision, or a failure to make a decision at all. It is sometimes theorized to be caused by analysis paralysis (over-analysis). Although such paradoxical situations might have been recognized by many of us for a long time, serious studies of the phenomenon seem not to have been seen in the United States and Western countries until recent publications, including Iyengar (Ph.D. dissertation, 1997) and Iyengar & Lepper (2000). Iyengar, Wells, and Schwartz (2006) studied how looking for the "best" job undermines satisfaction, and how we can sometimes find ourselves in the position of doing better but feeling worse. This paradoxical analysis was popularized by Barry Schwartz in his 2004 book, *The Paradox of Choice*. In the book, Schwartz argues that eliminating consumer choices can greatly reduce anxiety for shoppers. He stated: "Autonomy and freedom of choice are critical to our well being, and choice is critical to freedom and autonomy. Nonetheless, though modern Americans have more choice than any group of people ever had before, and thus, presumably, more freedom and autonomy, we don't seem to be benefiting from it psychologically."

In Chinese philosophy, it has been known for a long time that happiness is measured by the difference between what you desire and what you can get. Therefore, to be happier, one can either make efforts to get more or simply lower one's expectations. We introduced similar ideas as Paradoxes of Life, in the beginning of this book. *Tao Te Ching*, a well-known Chinese philosophical treatise, was written over 2500 years ago. It says about "choice" (Dale, 2002):

Color can make us blind!
Music can make us deaf!
Flavors can destroy our taste!
Possessions can close our options!
Racing can drive us mad and its rewards obstruct our peace!
Thus, the wise fill the inner gut rather than the eyes,
always sacrificing the superficial for the essential.

Nevertheless, in many situations, choice cannot be avoided. When this happens, the simple method described by Székely (1986) can assist you a great deal:

Suppose we are to choose one of n candidates (e.g., for marriage). If a "candidate" is not selected when it is his turn, then we will not have the opportunity to change our minds later. The question is, How can we select the best candidate if any of them can only be compared with the previous ones? If we always choose, e.g., the second one (or arbitrarily choose any one), the chance of selecting the best is $1/n$. Therefore, if the number of offers is great, the probability of selecting the best one is nearly 0. Surprisingly however, there is a method that enables us to select the best candidate with a probability of nearly 30%, even if n is a large number. Let the first 37%

(more precisely, $100/e\%$) of the candidates go and then select the first one better than any previous candidate (if none is better, select the last). In this case, the chance of selecting the best is approximately $1/e$ or 37%, regardless of the value of n.

PARADOX OF CHOICE

For those who are going to get married, I'd like to assign this homework: If the number of candidates n is uncertain or a random variable, what is your optimal strategy for selecting the best one?

1.4.8 *Paradox of Voting and Electing*

The United States has a federal government, in which the president is elected indirectly by the people, through an electoral college. In modern times, the electors virtually always vote with the popular vote of their state. For instance, if the Democratic candidate has a simple majority of the popular vote in Massachusetts, he/she will win all 12 electors in the state. The president elected will be the one who wins the most electors throughout the nation. The electoral college consists of the electors appointed by each state who formally elect the president and vice president of the United States. Since 1964, there have been 538 electors in each presidential election. Each state has a designated numbers of electors. For example, Alabama has 9 and California has 55 electoral votes.

The simple majority decision rule is rooted in the notion of individual fairness: one person one vote, whereas, the elector voting system was founded on the motives tempered by fairness norms that recognize both the entitlements of majority groups and the need to protect the interests and rights of minority groups. However, such a system creates controversies. Here is one from the perspective of statistic science.

A paradoxical situation created by this indirect voting system goes like this: a candidate who wins the popular vote will not win an election that is based on electoral votes. In the 1888 election, the winner of the popular vote did not become president. Democratic incumbent Grover Cleveland won the popular vote by a margin of 0.8% (90,596 out of 11,383,320 votes). Despite this slim popular victory, Republican Benjamin Harrison won the electoral college majority (233 out of 401 votes). The 2000 presidential election was the most recent election where the popular vote winner was not elected. George W. Bush ran on the Republican ticket against the Democratic candidate, the sitting vice president, Al Gore. Though Gore held a slim popular vote victory of 543,895 (0.5%), Bush won the electoral college 271,266, with one Gore elector abstaining (Wikipedia.org).

Not only the voting system, but also the *prediction* of the presidential election outcome is of great interest to news agencies and the public. However, due to the great uncertainties of outcomes in voting, predictions are often disappointing. The following famous ballot-theorem of W. A. Whitworth from 1878 (Székely, 1986) explains why.

Suppose there are two candidates, say, A and B, and that A scores n votes, B scores m votes, and $n > m$ (i.e., A wins). Then the probability that throughout the counting there are always more votes for A than for B is given by

$$p = \frac{n - m}{n + m}.$$

Thus, if A gets 60% and B gets 40% of the votes, i.e., $n = 1.5m$, then $p = 0.2$. That is, if A has received 50% more votes than B has, the probability that B had an equal or greater number of votes sometime during the counting is 4 times the probability that A was superior throughout the counting. This result is somewhat unexpected.

Another paradox related to voting is this: In a simple majority voting system, as soon as one alternative garners over 50% votes, the rest of the votes are irrelevant to the election result. But this can be applied to any single vote; thus, paradoxically, no vote is relevant to the election result. On the other hand, if we believe that the votes collectively determine the outcome, then it implies that the sum of nothingness can be significantly something.

1.4.9 *Brownian Motion Paradox*

Most of us probably remember hearing about Brownian motion in high school. But how many people know about Einstein's work on Brownian motion, which allowed Jean Perrin and other physicists to prove the physical reality of molecules and atoms? Brownian motion was one of the first crucial proofs of the discreteness of matter. First observed by Jan Ingenhousz in 1785, and later rediscovered by Brown in 1828, the phenomenon consists of the erratic or fluctuating motion of a small particle when it is embedded in a fluid (Parrondo and Dinıs, 2004). The quantitative explanation was given by A. Einstein 1905. Brownian motion is also called the Wiener process in recognition of Wiener's mathematical study of the Brownian motion. Brownian motion is commonly seen in our daily life. For instance, the movement of dusty particles in the air is Brownian motion in three-dimensional space. A random walk (imagine an inebriated person walking) can be considered as a two-dimensional Brownian motion.

Brownian motion provided a way to reconcile the paradox between two of the greatest contributions to physics at that time: thermodynamics and the kinetic theory of gases. Key to kinetic theory was the idea that the motion of individual particles obeyed perfectly reversible Newtonian mechanics. In other words there was no preferred direction of time. But the second law of thermodynamics expressly demanded that many processes be irreversible (Haw, 2005). Einstein's theory of Brownian motion explains why many processes are irreversible, in a probabilistic sense.

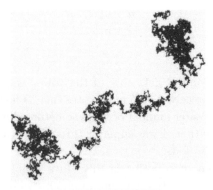

BROWNIAN MOTION

Brownian motion has some fantastic paradoxical properties, and here are three:

(1) Mathematically, Brownian motion is a motion with a continuous path that is nowhere differentiable with probability 1. This means that the instantaneous speed cannot be defined anywhere. This is because, according to Einstein's theory, the mean distance traveled by the particles is proportional to the square root of the time t. By taking the derivative of the mean distance with respect to t, the obtained mean speed is proportional to $1/\sqrt{t}$. From this it follows that the instantaneous ($t = 0$) speed of the particles would be infinite at any moment (Székely, 1986).

(2) The trajectories (realizations) of Brownian motion are rather irregular since they are nowhere differentiable. In the usual sense, we consider any irregular curve, such as the trajectory of planar Brownian motion, one dimensional. At the same time, it can be shown that the trajectory of a planar Brownian motion actually fills the whole plane (each point of the plane is approached with probability 1). Therefore, the trajectories can also be considered as two-dimensional curves.

(3) Brownian motion has the feature of self-similarity, i.e., the average features of the function do not change while zooming in, and we note that it zooms in quadratically faster horizontally (in time) than vertically (in displacement). More precisely, if $B(t)$ is a Brownian motion, then for every scaling $c > 0$ the process $\tilde{B}(t) = B(ct)/\sqrt{c}$ (the scaling law for fractal objects) is another Brownian process. The self-similarity property is also called scale invariance, and is the fundamental characteristic of fractals.

Brownian motion is memoryless motion; i.e., the future is only dependent on the present not the past. Brownian motion is recurrent in one or two dimensions (the particle returns almost surely to any fixed neighborhood of the origin infinitely often), whereas it is not recurrent in dimensions three and higher.

There are, increasingly, applications of Brownian motion in biology, ecology, game theory, the stock market, and drug development (Chang, 2010).

1.4.10 *Paradox of Random Graphs*

There are many interesting things we can learn from graph theory, for instance, the so-called Paradox of the Politician. Suppose in a group of people we have the situation that any pair of persons have precisely one common friend. Then there is always a person (the "politician") who is everybody's friend. You can find the mathematical proof from the book by Aigner and Ziegler (2004, p. 223).

In graph theory, there are two widely used random graph models developed by Erdős and Rényi (1959, 1960), denoted $G(n, M)$ and $G(n, p)$. In the

$G(n, M)$ model, a graph is chosen uniformly at random from the collection of all graphs that have n vertices (order n) and M edges (size M). For example, in the $G(3, 2)$ model, each of the three possible graphs on three vertices and two edges is included with probability $1/3$. In the $G(n, p)$ model, a graph is constructed by connecting nodes randomly. Each of the $\binom{n}{2}$ edges is included in the graph with probability p independent of the inclusion of other edges.

A large $(n \to \infty)$ random graph $G(n, p)$ has on average $\binom{n}{2}p$ edges (degrees) and the distribution of the degree k of any particular vertex is binomial, and approaches the Poisson distribution for large graphs if np is held constant, i.e.,

$$p(k) \approx \frac{npe^{-nk}}{k!}.$$

As $p(k)$ increases, one would expect that the graphical connectivity will increase gradually. However, according to Erdős and Rényi (1960), this intuition is misleading: Given an arbitrary positive constant $\varepsilon > 0$, if $p \leq \frac{(1+\varepsilon)\ln n}{n}$, then the random graph $G(n, p)$ will almost surely (with a probability of 1) contain isolated vertices, and thus be disconnected. On the other hand, if $p \geq \frac{(1+\varepsilon)\ln n}{n}$, then the random graph will almost surely be connected. Therefore, the probability $\frac{\ln n}{n}$ is a sharp divider for the connectedness of a random graph.

1.4.11 *A Ghost Number: Chaitin's Constant*

Chaitin's constant Ω in computer science is defined by the probability that a set of given instructions will halt on a universal Turing machine, a theoretical computer model invented by Alan Turing in the 1930s. The constant was constructed by Gregory Chaitin in 1976 to help determine whether a Turing machine will come to a halt given a particular input program. Chaitin's constant is a ghost number. No one knows its value or even one of its possible values, but we know what it represents. It is a Borel normal number that can be described but not computed; in other words, there is no set of formulas (as, for instance, the formulas for the calculation of e, π, or $\sqrt{2}$) or rules that when implemented by a program would allow one to compute it. Another interpretation of Ω is the following: If we repeatedly toss a coin, noting 1 if it turns up heads and 0 if it turns up tails, the probability of obtaining a sequence with length n representing a halting program is Ω_n. We can prove that Ω_n is algorithmically random, i.e., its complexity is n, by using the simple algorithm described, e.g., in the book by Marques (2008).

1.5   Chapter Review

We have provided various paradoxes from different fields to show you how powerful they can be in opening people's minds and sharpening their thinking. We showed you different views of democracy from the perspectives of process and outcome. We challenged the definition of identity through the paradox of identity switching, and promoted individual values through the paradox of leadership. We puzzled you with Simpson's Paradox, and raised an ethical issue via a paradox of murder. We logically disproved the existence of an omnipotent God using the Omnipotence Paradox. We surprised you with the smallness of the world we inhabit, containing us all within just about six degrees of separation.

We showed you the difference between music and noise in terms of wave structure, while pointing out the surprising fact that only 13 scales are used universally for music.

We showed you how misleading our intuition can be in the paradox of a fair game. We discussed the controversies about the use of DNA evidence in a justice system.

We have discussed various applications of paradoxes and have shown you how the counterintuitive Benford's law can be used to detect accounting fraud; how two losing games can be played sequentially to become a winning game and how this paradox can be used in stock diversification; and why, because of Arrow's Paradox, the ideal society cannot be achieved. We investigated the Friendship Paradox and its application in predicting disease outbreaks, applications of paradoxes in computer coding theory, the Inspection Paradox and its applications in machine replacement and genomics, and why the principle of an electric bell is actually based on a paradox.

We touched upon various well-known and consequential mathematical paradoxes. Achilles and the Tortoise, Hilbert's Grand Hotel, Torricelli's Trumpet of Finiteness and Infinity, and Berry's Paradox were postulated at different times, but all deal with infinity from different perspectives. The Barber Paradox has changed the face of set theory forever. The paradoxes of enrichment and the logistic mapping were used to introduce the two important concepts of bifurcation and chaos.

Fermat's last theorem, proved in 1995, was truly "the goose that lays the golden egg": The study of the theorem has generated many new mathematical branches.

Borel's normal numbers are stranger types of numbers that are defined by probability distributions, whereas paradoxes related to fixed points, the paradox of π, the Pretzel Transformation, and the Fractional Dimensions of

Fractals involve stranger geometric properties.

We also discussed several statistical paradoxes that do not involve in-depth reasoning or inference. The Roulette Paradox reveals the fact that there is a zero probability at any given point for a continuous probability distribution. The Birthday Paradox shows how a surprisingly small sample can provide an unexpected larger probability. The numbers of times an elevator goes up and down should, it seems, be approximately equal, as well as the average waiting time for going up or down. However, the Elevator Paradox explains how such an intuitive thought can be wrong. The Paradox of the Drunken Man is used to illustrate the concept of a mode in statistics. The Saint Petersburg Paradox discusses a decision theory problem with probability. The Paradox of the Family Tree addresses the probability of extinction in the branching process. The Paradox of Choice intimately connects to many common problems in our lives, and the simple 37% rule for decision-making is sometimes fairly useful. The Paradox of Voting and Electing is used to discuss the controversies in the presidential election with the electoral college, and explains statistically why it is so challenging to predict the outcome. Brownian motion is widely used in many different fields. In the Brownian Motion Paradox, we discussed several peculiar properties of this kind of motion. The paradox of Random Graphs discussed a chaotic connection property of a large random graph. A Ghost Number discusses Chaitin's constant, which is defined by the probability of halting for a given computer program.

I hope the paradoxical materials presented so far are powerful enough to keep your mind open and can motivate you to keep reading, to move on to the next topic.

In the next two chapters, we are going to discuss formal reasoning and inferential tools, i.e., probabilistic and statistical inference for scientific research, through various paradoxes.

1.6   Exercises

1.1 Discuss the following paradoxical statements. State whether you agree or disagree and why.

(1) A man who fears suffering is already suffering from what he fears.
(2) Happiness is the absence of striving for happiness.
(3) Reach the goal by giving up the attempt to reach it.
(4) If you want what you get, you will get what you want.
(5) Failure is the foundation of success, and success the lurking place of fail-

ure.

(6) The more they give, the greater their abundance.

(7) Shared joy is a double joy; shared sorrow is half a sorrow.

(8) Art is a lie that makes us realize the truth.

(9) I shut my eyes in order to see.

(10) Painting is easy when you don't know how, but very difficult when you do.

(11) The notes I handle no better than many pianists. But the pauses between the notes, ah, that is where the art resides.

(12) If infinities can be stored in and handled by a finite human brain, does it mean infinities can be smaller than finiteness?

(13) Probability is a measure of possibility, but probability 0 does not mean impossibility.

1.2 As we discussed in the Paradox of Democracy and Dictatorship, should we support dictatorship process-wise in order to achieve democracy quickly?

1.3 If you were Professor Lee in the Paradox of Identity Switching, would you be willing to switch? Would the decision be different if you were the student? Why or why not?

1.4 Does the electoral system in presidential elections ignore the Paradox of Leadership? How can we improve the system to more "realistically reflect" the people's will?

1.5 In the case presented in Paradox of the Murderer, how does the judicial system define murder in your country? What if a person had the intention to do one thing (e.g., saving a life or committing a murder), but the outcome was the opposite of what he intended?

1.6 In the Knowing-the-Fishing Bridge, the two people have an unclaimed common assumption. What is it? Do they assume that in order to know how a person feels or knows you have to be the person? Do you agree with this assumption?

1.7 Give examples of "What is Not" in your life? A child wants to know which family member is in the room, but she does not want to knock on the door since it may not be polite, so she looks at all the other rooms and finds out what she needs to know. How?

1.8 In the Omnipotence Paradox, do you think we have proved the absence of an omnipotent being? What is the unclaimed assumption in the paradox?

1.9 In the Paradox of A Fair Game, if the rules are:

(1) If head-head appears, A wins \$9 from B.

(2) If tail-tail appears, A wins x dollars from B.

(3) If head-tail or tail-head appears, B wins $5 from A.

Determine the amount x so that the game is truly fair.

1.10 Collect shopping receipts and examine the first digits of the payments. Does your finding support Benford's law?

1.11 Does Parrondo's Paradox convince you to diversify your stock? Why or why not?

1.12 Arrow's Paradox, the Paradox of Democracy and Dictatorship, the Paradox of Leadership, and a few other paradoxes in Chapter 1 concern sociology. Write a sociological essay including these paradoxes.

1.13 Give an (potential) application of the Friendship Paradox.

1.14 Give an (potential) application of the Inspection Paradox.

1.15 How does the story of Achilles and The Tortoise address the issues of the sum of infinite numbers of arbitrarily small numbers? How is the same issue addressed in the Roulette Paradox?

1.16 Write an essay on the role of the Barber's Paradox in developing modern set theory.

1.17 Torricelli's Trumpet has a finite volume and an infinite surface. Can you imagine an object that has a finite surface and infinite volume? If not, why?

1.18 According to the logic in Hilbert's Paradox of the Grand Hotel, assuming we can live forever, any two persons are equally rich as long as they continue earning some money; but how much they earn each day is irrelevant. Do you agree?

1.19 Can you find applications of the 37% rule proposed in the Paradox of Choice?

1.20 If in a presidential election, we make 10 predictions on a winner at a randomly selected time, what is the probability of one of the predictions being wrong?

1.21 In the St. Petersburg Paradox, how much are you willing to pay for the fixed amount to enter the game? Do you think any casino will produce such a game?

1.22 Can you think of any applications based on what you have learned from this chapter?

Chapter 2

Mathematical and Plausible Reasoning

Quick Study Guide

This chapter is the beginning of a serious discussion of paradoxes in scientific inference or reasoning. We will discuss two types of reasoning: formal logic and plausible reasoning.

We will begin with a review of the two different notions of probabilities: frequentist and Bayesian. We then unify them by examining the hidden assumption in the conception of Bayesian probability: causal space. We conceptualize causal space using examples in daily life and unify the two paradigms.

We will introduce the very basics of formal logic before studying probability and probabilistic inference. To facilitate our discussion, we will utilize some well-known paradoxes, including the Boy or Girl Paradox, the Coupon Collector's Problem, Bertrand's Paradox, the Monty Hall Dilemma, the Tennis Game Paradox, the Paradox of Nontransitive Dice, the Paradox of Ruining Time, and the Paradox of Independence. We will not only provide fresh views on classical paradoxes but also present new paradoxes.

2.1 Probability and Randomness

Probability is the most important concept in modern science, especially as nobody has the slightest notion what it means. – Bertrand Russell

My thesis, paradoxically, and a little provocatively, but nonetheless genuinely, is simply this: Probability does not exist.—B. de Finetti

In 1900, at the International Mathematical Congress in Paris, David Hilbert considered the problem of the foundation of probability theory as one of the 23 most important unsolved problems in mathematics. Nowadays it is widely accepted in most developed countries that a study of Statistics is an essential part of one's education. Since we are constantly surrounded both in everyday life and in our professional life by uncertainties, it is impossible to understand the world around us without being able to analyze such situations. Probability and Statistics provide the mathematical framework and tools for this task.

However, probability theory is infested with many misconceptions, fallacies, and pitfalls (Leviatan, 2002). Our mind, like the sense of sight, has its illusions. One of the great advantages of probability theory is that it teaches us to distrust our first impressions, whereas the study of paradoxes in probability and statistics can provide us with the necessary conceptual clarity.

2.1.1   *Conceptual Unification of Probability*

Probability is a simple concept to a non-statistician, but to a serious statistician probability is not a trivial concept. As the French artist Edgar Degas put it: "Painting is easy when you don't know how, but very difficult when you do."

Probably the simplest and well-known experiment about probability is coin-tossing. Buffon tossed a coin 4,040 times; heads appeared 2,048 times. K. Pearson tossed a coin 12,000 times and 24,000 times. Heads appeared 6,019 times and 12,012 times, respectively. For these three series of tosses the relative frequencies of heads (0.5049, 0.5016, and 0.5005) are examples of (empirical) probabilities. However, how do we define probability if the experiments can not be repeated, as when discussing the likelihood of rain tomorrow? In such cases, many people believe probability is subjectively defined as a measure of strength of belief. But what is this belief based on? Are there any relationships (common ground) between the two concepts of probability?

On the whole, there are two views of probability: (1) For a frequentist, an event's probability is the proportion of times that we expect the event to occur if the experiment were repeated a large number of times, or the ratio of the number of favorable outcomes and possible outcomes in a (symmetric) experiment; (2) for a Bayesian statistician, an event's probability is a reflection of our state of knowledge. A probability is a number between 0 and 1 assigned to an event. The following terms are often encountered in the discussion of probability.

- Experiment: A phenomenon where outcomes are uncertain, e.g., single throws of a six-sided die.
- Sample space: The set of all outcomes of the experiment, e.g., $S = \{1, 2, 3, 4, 5, 6\}$.

- Event: A collection of outcomes; a subset of S, e.g., $A = \{2\}$, $B = \{3,5\}$.

The Frequentist Notion of Probability

A frequentist defines an event's probability as the limit of its relative frequency in a large number of trials. This classical definition of probability is identified with the works of Pierre-Simon Laplace. The probability of an event is the ratio of the number of cases favorable to it to the number of all cases possible when nothing leads us to expect that any one of these cases should occur more than any other, which renders them, for us, equally possible. This definition is essentially a consequence of the *principle of indifference.* If elementary events are assigned equal probabilities, then the probability of a disjunction of elementary events is just the number of events in the disjunction divided by the total number of elementary events. Examples would be the coin-tossing experiments conducted by Buffon and Person.

The principle of indifference (principle of insufficient reason) is a rule for assigning epistemic probabilities. Suppose that there are $n > 1$ mutually exclusive and collectively exhaustive possibilities. The principle of indifference states that if the n possibilities are indistinguishable except for their names, then each possibility should be assigned a probability equal to $1/n$.

THE CONCEPT OF PROBABILITY

Bayesian Notion of Probability

Bayesian probability is one of the different interpretations of the concept of probability and belongs to the category of evidential probabilities. The Bayesian interpretation of probability can be seen as an extension of logic that enables reasoning with uncertain statements. To evaluate the probability of a hypothesis, the Bayesian probabilist specifies some prior probability, which can then be updated in the light of new relevant data. The Bayesian interpretation provides a standard set of procedures and formulas to perform this calculation. Bayesian probability interprets the concept of probability as "a measure of a state of knowledge."

Broadly speaking, there are two views on Bayesian probability that interpret the state of knowledge concept in different ways. According to the objectivist view, the rules of Bayesian statistics can be justified by requirements of rationality and consistency and interpreted as an extension of logic. According to the subjectivist view, the state of knowledge measures a "personal belief." Many modern machine learning methods are based on objectivist Bayesian principles. One of the crucial features of the Bayesian view is that a probability is assigned to a hypothesis, whereas under the frequentist view, a hypothesis is typically rejected or not rejected without directly assigning a probability.

The Dilemma

Bayesianism cannot be fully understood from the frequentist point of view, and vise versa. The fundamentals of Statistics, especially the many controversies between different paradigms, cannot be fully understood from the viewpoint of Statistics itself. It can only be understood profoundly from a broader view, i.e., from the general scientific point of view. To see the wood, not just individual trees, we have to step outside, and not view them from inside the forest.

Let's look into some everyday examples and how we should interpret the competing concepts in those situations. If frequentism and Bayesianism have different interpretations of probability, then how should we answer these questions when we are asked by a non-statistician?

(1) In tossing a fair coin, what is the probability the coin comes up heads? We may say that the probability of heads (or tails) is 50%. The classical concept of probability involves a long sequence of repetitions of a given situation.

(2) What is the probability of the Boston team winning the basketball game last night? The result of the game is given; what is the meaning of probability here?

(3) When predicting weather (raining or not) for tomorrow, it is often considered a random event. We may say, e.g., there is a 30% chance or probability of rain tomorrow. However, it is a one-time event, and there is nothing random about it. In fact, if we consider the time, the location, etc., every event is a single event and has nothing to do with randomness.

If the question is about the outcome of my next coinflip, the outcome is fixed but unknown to both you and me. If the question is about the outcome of my last coinflip, the outcome is fixed and known to me but unknown to you (if you didn't see me flipping the coin). Therefore, for any well-defined event, the result (either already known or unknown yet) is fixed; thus, there is no objective probability about the event. The concept of probability is really a description of the status of personal knowledge about event outcomes. Such knowledge is constructed based on the similarity principle: Similar process will likely result in the same outcome. For this reason, we conduct many identical experiments (coinflipping) under similar conditions, and use the proportion of heads to define the probability of "heads."

Conceptual Unification

To further investigate the difference in the conception of probability, from either the frequentist or Bayesian perspective, let me rephrase the questions in the above three probability examples, as follows:

(1) If I flip the same coin under similar conditions many times (N), what is the proportion of heads?
(2) If the Boston team plays many (N) times against the same team as last night, what is the proportion of wins for Boston?
(3) If the same (or similar) weather conditions as tonight appear many (N) nights in the next 1 billion years, what proportion of subsequent days are rainy?

If the answers to the three rephrased questions are essentially the same as those to the original questions, then we can say the probability concepts for the three situations are the same; i.e., the probability of an event is the ratio of the number of cases favorable to it to the number of all cases (N).

In the case of flipping a coin, we didn't specify who would perform the experiment, when it would be performed, how much force he/she will use, which coin will be used, etc. Instead, we used the general term "tossing a fair coin" and let each person imagine the experimental situation. For instance, some people may consider that a fair coin must have two identical faces, others may not. Such a personalized experimental condition leads to a personalized probability measure.

In guessing the winner of last night's basketball game, we probably evaluate the Boston team's relative skill against the opponent's. Such skill may be subconsciously estimated using the proportion of previously won games. The proportion may be weighted by the skill of each opponent (the skill may be measured by the proportion of wins); i.e., it is weighted more for winning against a highly skilled team than a less skilled one.

In assessing the probability of rain, we may recall in the past some of the same weather conditions similar to today's, and whether it rained or not the next day. Such predictive relationship is often vaguely formed (and called individual knowledge) before one is asked the weather question. It is fundamentally constructed based on the concept of the proportion of rainy days, even if complicated statistical models are used and the historical data is provided by others. It is exactly for this reason that people are not aware that their concept of probability is nothing but the proportion of favorable outcomes among the "similar situations" under consideration.

In general, in assessing a probability we always synthesize (use) similar situations and "average" their outcomes to form the probability. However, the "similar situations" are generally not specified, which leads to different considerations of similarities and, consequently, to a difference in probability calculations between frequentism and Bayesianism. The view of similarity is seen in a more restricted sense in the frequentist paradigm than in the Bayesian paradigm.

For convenience, we call the collection of all "similar situations" the causal space, whereas the collection of corresponding outcomes for situations in the causal space is defined as the outcome space. The outcome space is similar to but different from the sample space: Each outcome in the sample space has equal probability, but outcomes in an outcome space are not necessarily of equal probability because similar situations in a causal space do not necessarily have equal probabilities.

There is nothing truly identical, nor are experiments truly identical, in the world. If the experiments were really identical the results should be identical. For example, if all the conditions (excepting time) are the same in the "repeated" flipping coin experiments, you probably agree that there will be just one outcome, be it heads or tails. The reason for different outcomes is that other unknown factors affect the outcomes, such as a slight difference in forcing the coin or the smoothness of the surface where the coin drops. And because these factors may change in each experiment and affect the outcome, we consider them as random, and this randomness results in unpredictable outcomes in each experiment.

When we make a probability estimate, we implicitly apply the "similarity principle": Similar situations will likely result in the same outcome. Thus the results from these similar situations can be pooled together to form (implicitly) a probability distribution of the outcome. However, the method of pooling can be different for different statistical paradigms and even varies from person to person. In the Bayesian paradigm, the set of situations is often vaguely or implicitly defined; thus, different researchers will come up with very different values for the probability. In the frequentist paradigm, the set of situations (or experiments) is well defined, so the value of the prediction is usually unique. For this reason, Bayesianism is often criticized for its subjectivity in comparison to frequentism. However, this is a misconception; ignoring previous knowledge that is vaguely related to the current evidence, as frequentists do, will not make the conclusion more objective.

In the Bayesian paradigm, the subjective prior is formulated implicitly based on individual experience, knowledge, or belief. It argues that belief is from experience. A Bayesian subjective prior, in fact, uses the similarity principle and formulates a set of similar situations very vaguely and implicitly (intuitively or by instinct). In contrast, the objective prior will also use the similarity principle, but the "similar situations" are explicitly defined, and with a stricter sense of similarity (e.g., from a prior experiment) such that the calculation is possible and priors from different individuals can have some degree of agreement among different individuals. In a broader sense, scientific discovery is about identification of similarities, and these similarities can be either in the sense of "cause" or "consequence."

What is the probability of rain tomorrow? It seems a simple question, but if people understand probability differently, should the answer be dependent on the understanding of the questioner or the answerer? We usually expect the correct answer from an expert, as he gives an answer based on his concept of probability, but he might not be answering the question as it is framed in the mind of the questioner. What's the point then of having the expert's opinion?

In summary, whether frequentist or Bayesian, concepts of probability are based on the collection of similar phenomena or experiments. Frequentism explicitly defines the similarities or repeats the same experiment, whereas the Bayesian paradigm implicitly or semi-explicitly specifies the similarities. The implicitness of the Bayesian approach leads to a criticism that it is subjective.

There is only a single history, even though statisticians tend to think it as a random sample. If we believe everything exists or happens for a reason, then there is no randomness. If we believe that not all things happen for a reason, should we or can we determine which things happen for reasons?

We are now ready to delve deeper into different aspects of probability concepts through various paradoxes.

2.1.2   *The Boy or Girl Paradox*

Let's begin with a simple calculation of probability raised from the well-known Boy or Girl Paradox. The paradox dates back to at least 1959, when Martin Gardner published one of its earliest variants in the magazine *Scientific American*. He phrased the paradox in terms of two questions: (1) Mr. Jones has two children. The older child is a girl. What is the probability that both children are girls? (2) Mr. Smith has two children. At least one of them is a boy. What is the probability that both children are boys? Though the two questions may seem to be asking the same thing, the answers are different.

Recall that a sample space consists of all elemental events having the same probability, while the Principle of Indifference can be used to identify such elemental events. The principle states: "If there is no known reason for predicating of our subject one rather than another of several alternatives, then relatively to such knowledge the assertions of each of these alternatives have an equal probability" (Keynes, 1963, 42; Shackel, 2007).

Now denote the sample space by {BB, BG, GB, GG}, where we have labeled boys B and girls G and used the first letter to represent the older child. Gardner assumed that these outcomes are equally probable, and thus the answers to the above two questions are 1/2 and 1/3, respectively. The equal probability assumption implies:

(1) Each child is either male or female.
(2) The sex of each child is independent of the sex of the other.
(3) Each child has the same chance of being male as of being female.

The assumptions (1)—(3) have been shown empirically to be false (Nickerson, 2004), since we usually ignore the facts that a child could be intersex, the ratio of boys to girls is not exactly 50:50, and (among other factors) the possibility of identical twins means that sex determination is not entirely independent.

This paradox shows you how the sample space changes as the underlying question changes. The next example about coupon collection is a little more complicated. It involves the concept of joint probability.

2.1.3   *Coupon Collector's Problem*

Many of us collected sports cards (or coupons, or the like) in childhood. Stores often sell them in nontransparent envelopes, which helps them sell more cards. The reason is that if we want to collect all n different cards, we have to buy many more than n cards. It is obvious that it takes very little time to collect the first few specimens. On the other hand, it takes a long time to collect the last few. In fact, for $n = 50$ unique cards it takes on average 50 cards to purchases in order to collect the very last unique card, after the other 49 have been collected. This is why the expected number of cards needed to collect all 50 unique cards is much larger than 50.

In probability theory, such a problem is called a coupon collector's problem, and defined as follows: If one randomly draws cards from n cards with replacement, how many cards does one have to pick on average in order to obtain the n different cards.

If you already have drawn k distinct cards, then the probability of not getting a new one in the next drawing is $\frac{k}{n}$. So the probability of needing exactly s drawings to get the next new card is $\left(\frac{k}{n}\right)^{s-1}\left(1 - \frac{k}{n}\right)$. And thus the expected number of drawings for the next new card is

$$\sum_{s \geq 1} \left(\frac{k}{n}\right)^{s-1}\left(1 - \frac{k}{n}\right) = \frac{n}{n-k}.$$

So the expected number of drawings until we have drawn each of the n different cards at least once is

$$\sum_{k=0}^{n-1} \frac{n}{n-k} \approx n \ln n, \text{ for large } n.$$

When $n = 50$ it takes on average about 225 trials (cards) to collect all 50 different cards.

Let the random variable T_n be the number of trials needed to collect all n different cards. Then it can be proved (Aigner and Ziegler, 2004) that

$$P\left(T_n > n \ln n + cn\right) \leq e^{-c},$$

where $c > 0$ is a constant.

Coupon problems are encountered frequently in the real world. For instance, Gadrich and Ravid (2008) generalized the coupon collector's problem to the case of group drawings with indistinguishable items, and applied their results to a statistical quality control problem that arises in a dairy's bottle filling process with s nozzles, using sequential samplings.

2.1.4   *Bertrand's Paradox*

The term "random" often means "equal probability"; it is one of the most common words in Probability. However, inappropriate use of the term can easily cause great confusion. The Bertrand Paradox is a prominent example.

Produced by Joseph Bertrand in the nineteenth century, this paradox shows that probabilities may not be well defined if the mechanism that produces the random variable is not clearly defined. The paradox goes as follows. Consider an equilateral triangle inscribed in a circle. Suppose a chord of the circle is chosen at random. What is the probability that the chord is longer than a side of the triangle? Bertrand gave three arguments, all apparently valid, yet yield different results. The arguments follow (Clark, 2007).

The Random Endpoints Method: Choose two random points on the circumference of the circle and draw the chord joining them. The chord divides the circumference into 2 pieces, whereas the equilateral triangle equally divides circumference into 3 pieces. If the smaller piece of the former is larger than a piece from the latter, the chord is longer than a side of the triangle. Therefore, the probability that a random chord is longer than a side of the inscribed triangle is $1/3$.

The Random Radius Method: Randomly choose a radius of the circle and a point on the radius, then construct the chord through this point and perpendicular to the radius. To calculate the probability in question, imagine the triangle rotated so that a side is perpendicular to the radius. The chord is longer than a side of the triangle if the chosen point is nearer the center of the circle than the point where the side of the triangle intersects the radius. Because the side of the triangle bisects the radius, the probability that a random chord is longer than a side of the inscribed triangle is $1/2$.

The Random Midpoint Method: Randomly choose a point anywhere within the circle, and construct a chord with the chosen point as its midpoint. The chord is longer than a side of the inscribed triangle if the chosen point falls within a concentric circle of radius $1/2$ the radius of the larger circle. Because the area of the smaller circle is $1/4$ the area of the larger circle, the probability that a random chord is longer than a side of the inscribed triangle is $1/4$.

Other distributions can easily be imagined; for instance, Chiu and Larson (2009) derive five new possible interpretations of randomness.

2.1.5   *Monty Hall Dilemma*

Let's continue to explore the conceptualization of probability through a more complicated case, specifically, the Monty Hall dilemma (MHD). MHD was originally posed in a letter by Steve Selvin to the *American Statistician* in 1975 (Selvin 1975a, 1975b). A well-known statement of the problem was published in Marilyn vos Savant's "Ask Marilyn" column in the magazine *Parade* in 1990 (vos Savant, 1990).

Suppose you're on a game show and you're given the choice of three doors. Behind one door is a car; behind the others, goats. The car and the goats were placed randomly behind the doors before the show. The rules of the game show are as follows: After you have chosen a door, the door remains closed for the time being. The game show host, Monty Hall, who knows what is behind the doors, now has to open one of the two remaining doors, and he will open a door with a goat behind it. After Monty opens a door with a goat, he always offers you a chance to switch to the last, remaining door. Imagine that you chose Door 1 and the host opens Door 3, which has a goat. He then asks you "Do you want to switch to Door Number 2?" Is it to your advantage to change your choice?

Many readers refused to believe that switching is beneficial as Von Savant suggested. Among many thousand readers of *Parade*, there were nearly 1,000 with PhDs who wrote to the magazine claiming that vos Savant was wrong (Tierney 1991).

Ironically, Herbranson and Schroeder (2010) recently conducted experiments showing that stupid birds (like pigeons) can make the right decision when facing the Monty Hall Dilemma. They wrote: "A series of experiments investigated whether pigeons (Columba livia), like most humans, would fail to maximize their expected winnings in a version of the MHD. Birds completed multiple trials of a standard MHD, with the three response keys in an operant chamber serving as the three doors and access to mixed grain as the prize. Across experiments, the probability of gaining reinforcement for switching and staying was manipulated, and birds adjusted their probability of switching and staying to approximate the optimal strategy. Replication of the procedure with human participants showed that humans failed to adopt optimal strategies, even with extensive training."

The Monty Hall dilemma has attracted academic interest because the result is surprising and the problem is interesting to formulate. Furthermore, variations of the MHD are made by changing the implied assumptions, and the variations can have drastically different consequences.

As the player cannot be certain which of the two remaining unopened doors is the winning door, most people assume that each of these doors has an equal probability and conclude that switching does not matter. However, the answer may not be incorrect depending upon the host's behavior. You could increase the probability (p) of winning by switching from $1/3$ to $2/3$. Here is why. The player, having chosen a door, has a $1/3$ chance of having the car behind the chosen door and a $2/3$ chance that it's behind one of the other doors. It is assumed that when the host opens a door to reveal a goat, this action does not give the player any new information about what is behind the door he has chosen, so the probability of there being a car behind a different door remains $2/3$; therefore the probability of a car behind the remaining door must be $2/3$. Switching doors thus wins the car with a probability of $2/3$, so the player should always switch. However, the host's behavior might affect the probability and your decision. The following are some results depending on the host's behavior.

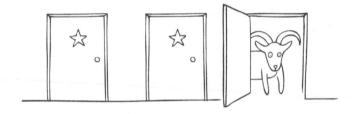

THE MONTY HALL PROBLEM

- If the host is determined to show you a goat, and he picks one at random with a choice of two, then $p = 2/3$.
- If the host is determined to show you a goat; with a choice of two goats (Bill and Nan, say), he shows you Bill with probability n, then given that you are shown Bill, the probability that you will win by switching doors is $p = 1/(1 + n)$.
- If the host opens a door chosen at random irrespective of what lies behind, then $p = 1/2$.

2.1.6   Tennis Game Paradox

Our intuition about probability can mislead us from time to time. Here is a very interesting example by Székely (1986). A family of three have decided to play tennis on the coming Sunday. The boy is going to play three tennis matches against his mother and father, and if he wins two matches in succession, he will be allowed to join his school's tennis team, as he had asked. He is offered a choice of two possible sequences of matches: father-mother-father or mother-father-mother. The boy has to decide which sequence will provide a larger winning probability knowing that his father plays better than his mother. The answer seems obvious: The second order is preferable for him as he plays twice with his mother in this version. However, a simple probability calculation shows our intuition is wrong.

Székely explains: If the boy wins against his father with probability p and with probability q against his mother, then $p < q$ since his father plays better than his mother. Choosing the first sequence the boy has to win either the first succession of two games (father-mother), the probability of which is pq, or the second succession (mother-father), the probability of which is qp. As a result, the probability that one of these two events will occur is $pq + qp - pqp$. If the boy chooses the second possible sequence then the probability he wins twice in succession is $qp + pq - qpq$. Since $p < q$, it follows that $pq + qp - pqp > qp + pq - qpq$, which means that it is preferable for the boy to choose the "father-mother-father" sequence. The difference between the two winning probabilities is $qpq - pqp$, which reaches it maximum value of 0.25 when the boy is as good as his father ($p = 0.5$) and definitely better than his mother ($q = 1$). A simple explanation for this paradox might be that in the father-mother-father sequence, the boy is offered two opportunities to beat his father (the one harder to beat), but in the other sequence, the boy has to win at the single opportunity against his father.

This paradox implies that it is sometimes easier to compete with a group of more skilled individuals than a group of less skilled individuals.

2.1.7 *Paradox of Nontransitive Dice: Round Defeat*

In mathematics, the well-ordering principle states that every non-empty set of positive integers contains a smallest element. We know that natural numbers are wellordered. However, the Paradox of Nontransitive Dice can surprise you completely (wikipedia.org; Savage, 1994).

A set of nontransitive dice is a set of dice for which the round-defeat relation holds similar to that in the game Rock, Paper, Scissors, in which each element has an advantage over one choice and a disadvantage to the other.

Efron's Dice

Efron's dice are the four dice A, B, C, D with the following numbers on their six faces:

- $A : \{4, 4, 4, 4, 0, 0\}$

- $B : \{3, 3, 3, 3, 3, 3\}$
- $C : \{6, 6, 2, 2, 2, 2\}$
- $D : \{5, 5, 5, 1, 1, 1\}$

In each game, one player chooses a die as he pleases, and then the other player can choose a die out of the remaining three. The dice are tossed, the player who gets a higher number wins. Surprisingly, you can afford to be a gentleman and let the other player always pick his die first, and yet make the game favorable to you! The reason is that die A beats die B; B beats C; C beats D, D beats A. As a result you can guarantee yourself a 2/3 chance of winning each game. That is,

$$P(A > B) = P(B > C) = P(C > D) = P(D > A) = 2/3.$$

B's value is constant; A beats it on 2/3 of the rolls because four of its six faces are higher. Similarly, B beats C with a 2/3 probability because only two of C's faces are higher. $P(C > D)$ can be calculated by summing conditional probabilities for two events:

- C rolls 6 (probability 2/6); wins regardless of D (probability 1)
- C rolls 2 (probability 4/6); wins only if D rolls 1 (probability 3/6)

The total probability of a win for C is therefore

$$P(C > D) = \frac{2}{6} \cdot 1 + \frac{4}{6} \cdot \frac{3}{6} = \frac{2}{3}.$$

Similarly, the probability of D winning over A is

$$P(D > A) = \frac{3}{6} \cdot 1 + \frac{3}{6} \cdot \frac{2}{6} = \frac{2}{3}.$$

The Efron Dice are optimal, in the sense that the lowest winning probability achieves the theoretical upper bound (Usiskin, 1964).

This paradox implies that it is not always the best strategy to make the first choice in life. For instance, you may not want to always choose a weapon before your opponent does!

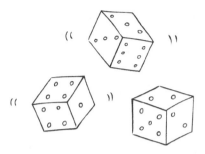

TWO-WAY ROUND-DEFEAT DICE

Two-Way Round-Defeat Dice

Similar to Efron's dice, there is a set of three dice (Székely, 1986; James Grime at http://singingbanana.com/dice/article.htm):

- $A : \{3, 3, 3, 3, 3, 6\}$,
- $B : \{2, 2, 2, 5, 5, 5\}$,
- $C : \{1, 4, 4, 4, 4, 4\}$,

which have a constant round-defeat probability of $7/12$; that is,

$$P(A > B) = P(B > C) = P(C > A) = 7/12.$$

Another amazing fact about this set of dice is the following: Suppose each player rolls twice the same die he/she has chosen, and that the player with the highest total wins. One may expect that the player who chooses later will double his/her chances of winning. Surprisingly, with two dice the order of the winning chain flips! That is,

$$P(A > C) = 671/1296, \ P(C > B) = 765/1296, \ P(B > A) = 765/1296.$$

If knowing non-transitive dice can help you win the game and impress your friends, the next paradox will help you on a trip to a casino.

2.1.8 *Paradox of Ruining Time*

Suppose you have initially a capital of m dollars, and decide to play a game with a probability p of winning and $q = 1 - p$ of losing. Each time you bet one dollar. Assume you have decided to keep playing until you are bankrupt or win $n > 0$ dollars from your opponent, who has at least n dollars. This problem can be viewed as a one-dimensional Random Walk (Brownian motion). Jacob

Bernoulli demonstrated in 1680 that such a game ends in a finite number of steps with a winning probability P that is a function of the initial capital m (Taylor and Karlin, 1998, p. 144—154):

$$P(m) = \frac{1 - (q/p)^m}{1 - (q/p)^{m+n}}, \ p \neq 0.5.$$

For instance, for $p = 0.48$, if you have an initial value of $m = \$100$ and want to win another $n = \$100$ with \$1 bets, the winning probability is $P(m) = P(100) = 0.03\%$. Better keep this in mind when you're planning trip to Las Vegas.

It can be easily seen that when $p = 0.5$, $P(m) = m/(m+n)$. In this case, if two players begin with m and n dollars, how many rounds of the game do they have to play on average before one of them is ruined? The answer is mn (see also Exercise 2.9). This result is surprising to us since, for example, if you have \$1 and your opponent has \$100, one hundred rounds are expected to be played on average before the game is over!

2.1.9 *Paradoxes of Independence*

A fundamental concept in probability is independence of events. Two events are independent if the occurrence of one is not dependent on the other. However, there is a distinction between pairwise independence and mutual independence.

A set of n events is said mutually independent if for any number k ($\leq n$) of these events, $A_1, A_2, ..., A_k$, the multiplication rule

$$P(A_1 A_2 \cdots A_k) = P(A_1) P(A_2) \cdots P(A_k)$$

holds. If the equation only holds for $k = 2$, then the n events are said to be pairwise independent.

Székely (1986, p. 13) illustrate that mutual independence is more than pairwise independence through the following interesting example:

Suppose one tosses two fair coins. Denote by A the event "the first coin falls heads up," by B the event "the second coin falls heads up," and by C the event "one (and only one) of the coins falls heads up." Then, surprisingly, the events A, B and C, are pairwise independent, but any two of the three uniquely determine the third.

It is obvious that A and B are independent. The events A and C (and B and C), however, do not seem to be independent at first sight; but, since

$P(AC) = P(A)P(C) = \frac{1}{4}$ (and similarly $P(BC) = P(B)P(C)$), they are really independent. It is also true that any two of the events determine the third because each event (A, B or C) occurs exactly when one and only one of the other two events occurs. Similar examples are given by Bohlmann (1908), Bernstein (1928), and Stoyanov (1997, p. 13).

One may now think if any three events satisfy the equation $P(ABC) = P(A)P(B)P(C)$, they must be mutually independent. Again, surprisingly, it is not necessarily the case, as illustrated by Stoyannov (1997, p. 14):

Suppose we toss a fair die with eight sides labeled 1 through 8. Consider $A_1 = \{1,2,3,4\}$, $A_2 = A_3 = \{1,5,6,7\}$, and $A_1 A_2 A_3 = \{1\}$. Then $P(A_1) = P(A_2) = P(A_3) = \frac{1}{2}$ and $P(A_1 A_2 A_3) = \frac{1}{8} = \frac{1}{2} \cdot \frac{1}{2} \cdot \frac{1}{2} = P(A_1)P(A_2)P(A_3)$. However, A_1 and A_2 are identical (not independent). Therefore, the three events are not mutually independent.

We know that if random variables X and Y are independent, then, for any two continuous real functions g and h, $g(X)$ and $h(Y)$ are also independent. The converse statement is not true unless g and h are 1-1 mappings. If the 1-1 condition does not hold, then $g(X)$ and $h(Y)$ can be independent while X and Y are not. Stoyanov (1997, p. 58) provide examples in which X_1 and X_2 are independent random variables, but X_1^2 and X_2^2 are dependent.

Conversely, if g and h are two nonconstant real functions of the same random variables, we would not expect them to be independent. However, Székely (1986) provides a surprising example: If X_1 and X_2 represent the co-ordinates of the velocity vector of a particle moving randomly in a plane and X_1 and X_2 are independent standard normal variables, then the quantities $Y = X_1^2 + X_2^2$ (which is proportional to the kinetic energy) and $Z = X_1/X_2$ (which determines the direction of the motion) are independent. The interpretation of this paradox is that the particle's energy (or magnitude of its velocity) is the same in all directions, i.e., is independent of orientation.

When the correlation of two random variables, X_1 and X_2, is zero, i.e., $E(X_1 X_2) = 0$, we say the two variables are uncorrelated. Random variables can be uncorrelated but not independent. For instance, if θ is uniformly distributed on $(0, 2\pi)$, then $X_1 = \sin\theta)$ and $X_2 = \cos\theta$ are uncorrelated since

$$E(X_1) = 0,\ E(X_2) = 0,\ \text{and } E(X_1 X_2) = 0.$$

However, X_1 and X_2 are not independent because they are functionally dependent: $X_1^2 + X_2^2 = 1$. A similar example given by Stoyanov (1997, p. 64) is: Suppose $X_1 \sim N(0,1)$ and $X_2 = X_1^2 - 1$. Then $E(X_2) = 0$ and $E(X_1 X_2) = 0$. Hence, X_1 and X_2 are uncorrelated. But they are functionally independent.

2.1.10   *Randomness and Complexity*

Probability is a study of randomness, but what is randomness anyway? We have never given a formal definition of randomness. How can one judge the randomness of a sequence of symbols? The Russian mathematicians Kolmogorov and Usphenski proposed the following criteria to examine sequence randomness (Marques, 2008):

- Typicality: The sequence must belong to a majority of sequences that do not exhibit obvious patterns.
- Chaoticity: There is no simple law governing the evolution of the sequence's terms.
- Frequency stability: Equal-length subsequences must exhibit the same frequency of occurrence of the different symbols (for a uniform distribution).

It is an interesting fact that the human specification of random sequences is a difficult task. This is because, paradoxically, any sequence can be from a random sequence or from a nonrandom sequence.

Kolmogorov's *complexity* is another definition of randomness. According to Kolmogorov and Chaitin, a series of numbers is random if the smallest algorithm capable of specifying it in a computer has about the same number of bits of information as the series itself. For instance, the series 0101010101010101 is not random, since it can be specified by saying "print 01 eight times." If the sequence was longer, then one would only have to alter the formulation by saying, for instance, "print 01 a thousand or a million times." Thus, one could say that the information in the sequence 0101. ..01 (a million times) is compressible (Byers, 2007, p. 325).

What if the information in the sequence is incompressible? What if the most efficient way to specify the digits in the sequence is to actually list them? Then we say that the sequence is random.

The "incompressibility" approach to randomness is really a very natural one. Chaitin refers to Ray Solomonoff's model of scientific induction, in which a scientific theory is seen as a compression of the data obtained through a scientist's observations. The "best" scientific theory would be the minimal program that could generate the data (consistent with Occam's Razor in Chapter 4). A "random" theory in this sense would be no improvement on the actual raw data. Using the algorithmic approach to randomness has led to the development of a tool for measuring randomness. This involves the concept of complexity. The complexity of a series of digits is the number of bits that must be put into a computer in order to obtain the original series as output, that is, it is the size (in bits) of the minimal program for the series. Thus, a random series of digits is one whose complexity is approximately equal to its size (Byers, 2007).

The algorithmic approach is fascinating and has a number of very deep implications. The first is that, whereas one can show that most numbers are random, we can never prove whether an individual number is or is not random. This last result is in the spirit of Gödel's incompleteness theorem. The argument is, like Gödel's, based on a paradox, in this case Berry's Paradox. Also, like Gödel's theorem, Chaitin's work has implications for the philosophy of mathematics. In particular, it puts restrictions on the information that can be derived from a formal axiomatic system.

In my view, any individual sequence cannot be proven random or not, because the notion of randomness is an attribute of a group not an individual.

2.1.11 *Randomness and Disorder*

Randomness is closely related to disorder, which is measured by entropy in information. The term *entropy* was coined for the first time in 1865 by the German physicist Rudolf Clausius in relation to the second principle of thermodynamics. Later, in 1877, the Austrian physicist Ludwig Boltzmann derived a formula $S = k \log n$ for the entropy of an isolated system, where n is possible states at the molecular level (microstates) needed to describe a macroscopic system, and k is a constant. The similarity is noticeable between the physical and mathematical entropy formulas (Marques, 2008, p. 181).

Suppose two types of gases in a large container are initially separated by a divider so that one gas is in each half of the container. After the divider is removed, the molecules of gases from different sides start to randomly travel inside the whole container. As the time goes by, the molecules of the two gases are mixed and almost uniformly mixed at equilibrium.

The entropy evolution in this case is a general phenomenon that we would observe in large isolated physical systems. The question is, would it be possible at certain time point later all molecules of one kind to be located in one side of the container and the other kind of molecules to move the other side, i.e., for the two gases to be completely separated again? Yes, it is theoretically possible, but the probability is virtually zero. It is this tendency-to-disorder phenomenon and its implied increase of entropy in an isolated system that is postulated by the famous second principle of thermodynamics. The entropy increase in many isolated physical systems is an indicator of time's "arrow". However, in reality virtually all systems are open, and there are "forces" that push the system in the direction of orderliness (Marques, 2008). An example in social science would be that dictatorship is the most ordered state and democracy is the most disordered; the world is moving from dictatorship to democracy, i.e., from order to disorder, with increasing entropy.

Boltzmann's formula is applicable for perfect randomness or disorder; i.e., a uniform probability distribution exists for each state. The entropy for the general situation is

$$H(K) = -\sum_{k=1}^{K} P(S_k) \log_2 P(S_k),$$

where the K states are denoted by $S_1, ..., S_K$ and $P(S_k)$ is the probability associated with state S_k.

To illustrate the formulation, Marques (2008) investigate various algorithms to generate English texts that contain only lowercase letters and the space character (27-symbol set):

The zero-order model: randomly selecting a character based on the uniform distribution $P(S_k) \equiv 1/27$, we obtain the maximum possible entropy for our random-text system, i.e., $\log_2 27 = 4.75$ bits. An example of an 85-character text produced by the zero-order model is

"rlvezsfipc   icrk-
bxknwk awlirdjokmgnmxdlfaskjn jmxckjvar jkzwhkuulsk odrzguqtjf
hrlywwn"

No similarity with a real text written in English is apparent.

The first-order model uses the probability distribution of the characters in true English texts. For instance, the actual rate of occurrence for the letter e is about 10.2% or $P(e) = 0.102$; the letter z has 0.04% of occurrences, so $P(z) = 0.0004$, etc. Using this first-order model, the entropy decreases to 4.1 bits, reflecting a lesser degree of disorder. The second-order model is to impose true probabilistic rules between two conjunctive letters in English. The entropy reduces to 3.38 bits. The move to third- and fourth-order models introduces further restrictions on the distribution. For the fourth-order model with an entropy of 2.06 bits, here is an example of text:

"om some people one mes a love by today feethe del purestiling touch one from hat giving up imalso tRe dren you migrant for mists of name dairst on drammy arem nights abouthey is the the dea of envire name a collowly was copicawber beliver aken the days dioxide of quick lottish on diffica big tern revolvetechnology as in a country laugh had"

For a fifth-order model there are 275 possible 5-letter sequences, that is, more than 14 million combinations, although the so-called Zipf law can be used as an approximation. The American linguist George Zipf studied the statistics of word occurrence in several languages. Let n denote the rank value of the word based on its occurrences (the second most commonly used English word has rank 2, for example) and $P(n)$ the corresponding occurrence probability. Then Zipf's law for English prescribes the following dependency of $P(n)$ on n:

$$P(n) = \frac{0.08}{n^{0.96}}.$$

Using this formula, it was estimated that for English texts the measure of entropy is 1.26 bits per letter. The value of entropy varies among languages (Marques, 2008; wikipedia), as one might surmise.

Paradoxically, a study from Cambridge University in United Kingdom indicates that the order of letters within words seems not important, i.e., different orders deliver the same information or meaning:

AOCCDRNIG TO RSCHEEARCH AT CMABRIGDE UINERVTISY, IT DEOSN'T MTTAER

IN WAHT OREDR THE LTTEERS IN A WROD ARE, THE OLNY IPRMOATNT TIHNG

IS TAHT THE FRIST AND LSAT LTTEER BE AT THE RIGHT PCLAE, THE RSET

CAN BE A TOATL MSES AND YOU CAN SITLL RAED IT. TIHS IS BCUSEAE THE

HUAMN MNID DEOS NOT RAED ERVEY LTETER BY ISTLEF, BUT THE WROD

AS A WLOHE.

2.1.12 Quantum Probability

Unlike the Laplacian concept of the deterministic law of nature, for quantum phenomena the probabilistic description is not just a handy trick, but an absolute necessity imposed by their intrinsically random nature. Quantum phenomena are "pure chance" phenomena. Quantum mechanics thus made the first radical shift away from the classical concept that everything in the universe is perfectly determined. However, since quantum phenomena are mainly related to elementary particles, i.e., microscopic particles, the conviction that, in the case of macroscopic phenomena (such as tossing a coin or a die) the probabilistic description was merely a handy way of looking at things and that it would in principle be possible to forecast any individual outcome persisted for quite some time. In the middle of the twentieth century several researchers showed that this concept would also have to be revised. It was observed that in many phenomena governed by perfectly deterministic laws, there was in many circumstances such a high sensitivity to initial conditions that one would never be able to guarantee the same outcome by properly adjusting those initial conditions. However small a deviation from the initial conditions might be, it could sometimes lead to an unpredictably large deviation in the experimental outcomes. The only way of avoiding such chaotic behavior would be to define the initial conditions with infinite accuracy, which amounts to an impossibility. Moreover, it was observed that much of this behavior, highly sensitive to initial conditions, gave birth to sequences of values—called chaotic orbits—with no detectable periodic pattern. It is precisely this chaotic behavior that characterizes a vast number of chance events in nature (Marques, 2008). We can thus say that randomness has a threefold nature:

(1) A consequence of our incapacity or inability to take into account all factors that influence the outcome of a given phenomenon, even though the applicable laws are purely deterministic. This is the coin tossing scenario.
(2) A consequence of the infinite sensitivity of many phenomena to certain initial conditions, rendering it impossible to forecast future outcomes although we know that the applicable laws are purely deterministic. This is the nature of deterministic chaos, which is famously manifest in our planet's weather and many, many other dynamical systems.
(3) The random nature of phenomena involving microscopic particles, which are described by quantum mechanics. Such a system is infinitely sensitive to small disturbances, making measurement impossible without disturbing the system. The best one can do is to determine the system statistically.

The conceptual discussion of probability serves as a basis for plausible reasoning, since we are often dealing with uncertain events. Having said that, it is a good idea to study formal reasoning before plausible reasoning since formal reasoning is also used as a part of plausible reasoning.

2.2   Mathematical Logic and Formal Reasoning

Mathematical logic and formal reasoning are important tools in scientific discovery and inference. Everyone, even a child, has some experience of using formal reasoning, even if only subconsciously. Here, we are going to introduce some very elemental concepts and laws. We frequently use these laws in daily life, and in scientific research, surely, but perhaps not in a mathematical form. Formal reasoning is especially heavily used in scientific philosophy. We will see how some of these mathematical laws can be challenged on a more fundamental basis. We will use the elementary laws in Fitch's Paradox of Knowability, Gödel's incompleteness theorem, and in this book's final section, which looks at artificial intelligence. However, you can skip this section for now if you prefer.

Induction

Induction, or inductive reasoning, is the process by which generalizations are made based on individual instances. However, Mathematical Induction is not a form of inductive reasoning. Mathematical Induction is a form of deductive reasoning.

Deduction

Deductive reasoning is reasoning whose conclusions are intended to necessarily follow from its premises. Deductive reasoning "merely" reveals the implications of propositions, laws, or general principles, so that, as some philosophers claim, it does not add up to truth. Deductive reasoning applies general principles to reach specific conclusions, whereas inductive reasoning examines specific information, perhaps many pieces of specific information, to impute a general principle.

Causal Inference

A causal inference draws a conclusion about a causal connection based on the conditions of the occurrence of an effect. Premises about the correlation of two things can indicate a causal relationship between them, but additional factors must be confirmed to establish the exact form of the causal relationship (wikipedia.org).

Before we study plausible reasoning, let's review the basics of mathematical reasoning, including some propositional calculus, the elementary laws in proposition calculus, and Mathematical Induction. We follow the syntax convention in the *CRC Standard Mathematical Tables and Formulae* by Daniel Zwillinger (2003).

With propositional calculus, we will see how easy it is to convert a formal reasoning problem into an arithmetic problem so that it can be handled by a computer as has been done in the proof of the famous four-color problem.

2.2.1 *Propositional Calculus*

A proposition is the natural reflection of a fact, expressed as a sentence in a natural or artificial language. Every proposition is considered to be true (T) or false (F). This is the principle of two-valuedness.

Truth Table of Propositional Calculus

		Conjunction	Disjunction	Implication	Equivalence	Negation
A	B	$A \wedge B$	$A \vee B$	$A \longrightarrow B$	$A \longleftrightarrow B$	$\neg A$
T	T	T	T	T	T	F
T	F	F	T	F	F	F
F	T	F	T	T	F	T
F	F	F	F	T	T	T

Note: T = True, F=False.

In mathematical logic, propositional calculus is the study of statements, also called propositions, and compositions of statements. A system of inference rules and axioms allows certain formulas to be derived, called theorems. The statements are combined by means of propositional connectives such as *and* ($\vee$), *or* ($\wedge$), *not* ($\neg$), *implies* ($\longrightarrow$), and *if and only if* ($\longleftrightarrow$, equivalence). Propositions are denoted by $\{A, B, C...\}$. The precedences of logic operators (connectives) are: parentheses, $\neg$, $\wedge$, $\vee$, $\longrightarrow$, $\longleftrightarrow$. Sometimes, we omit $\wedge$ for simplicity. The conjunction, disjunction, implication, equivalence, and negation in relation to their component element or events are self-explanatory, and there are summarized in the table. However, I'd like to emphasize an important point here: When A is false, then $A \longrightarrow B$ is true, regardless of B. The implication of this is that *Truth* is what we cannot prove to be wrong, and so is not just limited to what can be proved to be correct. For instance, suppose you claim that you would have won the lottery if you had bought a lottery ticket. This claim cannot be proved wrong (also cannot be proved correct), since you didn't buy the ticket; therefore, the claim is valid or true.

To simplify calculations, the logic value (true or false) of a statement can be calculated using logic functions. Let the value for true $= 1$ and false $= 0$, then we can perform a calculation representing any logical expression formed from elementary statements and connectives by using a computer. For instance, the arithmetic function AB corresponds to the connective (compound statement) $A \wedge B$, the function $A+B-AB$ to $A \vee B$, $1-A+AB$ to $A \longrightarrow B$, $1-A-B+2AB$ to $A \longleftrightarrow B$, and $1 - A$ to $\neg A$.

2.2.2 *Elementary Laws in Propositional Calculus*

Many laws that we use consciously or subconsciously in our daily lives, and in scientific research, can be expressed in axioms or by mathematical tautologies. A tautology is a statement which always is true. Some of the most commonly used elementary laws in propositional calculus are

- The excluded middle law: $A \vee \neg A$. It says, e.g., that one of the two statements must be true: "You have the book." or "You do not have the book."
- Modus ponens: $((A \longrightarrow B) \wedge A) \longrightarrow B$. It says, e.g., that if you have the book, you must have read it; you have the book; therefore, you have read it.
- Reductio ad absurdum (proof by contradiction): $(\neg A \longrightarrow A) \longrightarrow A$. If says that if the negation of A implies A, then A is true. The notion behind this law is the consistency axiom, which asserts that A and $\neg A$ cannot coexist in a system.

These elementary laws in propositional calculus can be easily verified using the corresponding arithmetic functions described in Section 2.2.1. For instance, the excluded middle law, $A \vee \neg A \equiv 1$, can be proved as follows:

$$A \vee \neg A = A + (1 - A) - A(1 - A) = 1 - A(1 - A) \equiv 1.$$

The law of modus ponens is verified in this way:

$$((A \longrightarrow B) \wedge A) \longrightarrow B = 1 - (1 - A + AB)A + (1 - A + AB)AB \equiv 1.$$

Paradox of Truth

The consistency axiom, $A \wedge \neg A \equiv F$, says that statement A and its negation $\neg A$ cannot be correct at the same time. Thus, the excluded middle law and the consistency axiom ensure that one and only one of the two statements must be true: "You have this book" or "You don't have this book." This notion seems so universally accepted that we often subconsciously use it. Surprisingly, the following paradox would make you think twice about its validity:

A patient's recovery from disease may sometimes depend on what the doctor informs him. If the doctor says: "You will recover soon," the patient could recover soon due to a psychological effect. On the other hand, if the doctor says: "You will not recover soon," the patient might not recover soon for the same reason. Therefore, the statements A and $\neg A$ are both true—the doctor is always right. In Chapter 4, I will challenge Reductio ad Absurdum by way of the pigeonhole principle.

2.2.3 *Predicate Calculus*

For developing the logical foundations of mathematics, we need a logic that has a stronger expressive power than propositional calculus, and to describe the properties of most objects in mathematics and the relations among them a richer language is needed. Predicate calculus is the language in which most mathematical reasoning takes place. Predicates are variable statements that may be true or false, depending on the values of their variables. Predicate calculus uses the universal quantifier $\forall$, the existential quantifier $\exists$, predicates $P(x)$, $Q(x,y)$, variables $x, y, ...$, and assumes a universe U from which the variables are taken. For instance, $P(x) : x > 10$, $\exists x \in U \; P(x)$ is a shorthand of $\exists x \; (x \in X \wedge P(x))$, which stands for "there exists a real number x such that $x > 10$." For $Q(x) : x^2 = 1$, $\forall x \in X \; Q(x)$ is shorthand for $\forall x (x \in X \longrightarrow Q(x)$, which means "for all real numbers x, $x^2 = 1$."

2.2.4 *Mathematical Induction*

Mathematical induction is a widely used tool of deductive (not inductive) reasoning. Mathematical induction is a method of mathematical proof typically used to establish that a given statement is true for all (or some) natural numbers, following these steps of proof:

(1) The **basis (base case)**: Providing the statement to be proved is a function of the natural number n, prove the statement is true for a natural number n_0.
(2) The **inductive step**: Assuming the statement is true for any natural number $n \geq n_0$, prove the statement is true for $n + 1$.

Example: Arithmetic Progression

Prove the following identity holds for any positive integer n:

$$1 + 2 + \cdots + n = \frac{n(n+1)}{2}. \tag{2.1}$$

Basis: It is obvious that for $n = 1$ the above equation holds.
Inductive Step: If the above equation holds for n, we prove that it holds for $n + 1$, that is

$$(1 + 2 + \cdots + n) + (n + 1) = \frac{(n+1)[(n+1)+1]}{2}. \tag{2.2}$$

This can be done easily if we substitute (2.1) into the left-hand side of (2.2) and simplify.

Mathematical Induction has been generalized to a structural induction in mathematical logic and computer science, where the recursion is applied to structures beyond the natural numbers, such as structural trees (wikipedia.org). Another form of Mathematical Induction is the so-called complete induction, in which the second step is modified to assume not only that the statement holds for n, but also for all m satisfying $n_0 \leq m \leq n$.

In scientific reasoning, we are often attracted by things that appear to be attractive, and intentionally or unintentionally neglect other things which otherwise are informative. The Three-Hat Puzzle illustrates how silence can be quite informative!

2.2.5 *Significance of Silence*

In today's world, our lives are strongly influenced by product recommendations from professional critics and experts that are available from numerous sources: television, magazines, radio, Internet, and so on. Very often these recommendations shape our decisions and choices (Kamakura et al., 2006). From the viewpoint of Attention Economics, a stentorian voice is a way to maximize your audience. But what about the statement "Silence is golden?" What is the meaning of silence, anyway?

Silence is the tranquility of the inner life, the quiet at the depths of its hidden streams. It is when the soul abandons the restlessness of purposeful activity. It is not so much a lack of sound as it is a cultivation of interior stillness. When we reflect on how many times we have seen a child totally engrossed in drawing, we get our first beautiful picture of silence. Two persons in love, blissfully gazing at each other, portray another profound quality of silence (Hemrick, 1999). In Greek, the word calm means "the heat of the day." It signifies a resting place at high noon, a spot preferable placid, peaceful, and cool. But how can we feel silence in such a noisy surrounding? The solution is simple: You hitch a ride on the wave of the sound.

Silence can be meaningful even in solving logic puzzles. Here is an example.

THE HAT PUZZLE

The Three Hat Puzzle

The game started in a completely dark room with five hats (3 red and 2 black, as everyone knew), when three equally talented players, John, Mike, and Bill, each picked a hat to put on. (In fact, they all took red hats this time.) After walking out of the room, each could only see the other players' hats and began thinking of how to guess the color of his own.

For a moment, no one spoke. At the second moment, there was a silence again. However, at the third moment they all knew that it was the red hat they were wearing. Why? Did the two silences tell the players something? Here is what transpired after each player saw his friends' red hats.

At the first moment, John reasoned: There are two black hats. My hat could be a black one. Mike and Bill reasoned similarly.

At the second moment, John reasoned: If someone saw the other two wearing black hats, he would have known that he was wearing red, since there were only two black hats. Because at the first moment no one spoke, it meant there was at most one black among them. Mike and Bill reasoned the same way.

At the third moment, John thought: If my hat is black, Mike and Bill would have shouted out a moment earlier that they had red hats, since they have already concluded there is at most one black hat among us all. They didn't speak at the second moment, and thus John came to the conclusion that his hat was red. Mike and Bill figured out their colors in the same way.

In this case, the players did use the two silences to correctly determine their answers.

2.3   Plausible Reasoning

Plausible reasoning is reasoning involving probability. A simple example:

> If A is true, then B becomes more plausible.
> B is true.
> _____
> Therefore, A becomes more plausible.

Another example would be:

> If A is true, then B becomes more plausible.
> A becomes more plausible.
> _____
> Therefore, B becomes less plausible.

A prediction draws a conclusion about a future individual from a past sample. Proportion Q of observed members of group G have had attribute A. Therefore, there is a probability corresponding to Q that other members of group G will have attribute A when next observed. Here the group G is what we called earlier a "collection of similarities."

2.3.1 *Bayesian Reasoning*

Bayes' rule plays a fundamental role in Bayesian reasoning. If A and B are two random events, then the joint probability of A and B, due to symmetry, can be written as $P(B|A) P(A) = P(A|B) P(B)$, where $P(X|Y)$ denotes the conditional probability of X given Y. From this formulation, we can obtain Bayes' rule:

$$P(B|A) = \frac{P(A|B) P(B)}{P(A)}.$$

Bayes' rule is useful when $P(A|B)$ is known and $P(B|A)$ is unknown. Bayes' rule is often converted into different forms in practice. Let D be the observations and H_i be hypotheses ($i = 1, 2, ..., n$) with one and only one H_i being true. The probability of data D happening is given by the law of total probability:

$$P(D) = P(D|H_1) P(H_1) + \cdots + P(D|H_n) P(H_n).$$

Assume, in addition, that the probabilities of hypotheses $H_1, ..., H_n$ are known (prior probabilities). Then the conditional (posterior) probability of the hypothesis H_i, given by that data D, is

$$P(H_i|D) = \frac{P(D|H_i) P(H_i)}{P(D)}.$$

This is another of form of Bayes' rule.

DNA Evidence in A Courtroom

Bayesian reasoning can be used in a court setting by a juror to coherently accumulate evidence for and against the guilt of the defendant, and to see whether, in totality, it meets his or her personal threshold for "beyond a reasonable doubt" (Dawid and Mortera, 1996; Foreman, Smith, and Evett, 1997; Robertson and Vignaux, 1995). To illustrate this, we denote by

- H_0 the event that the defendant is not guilty.
- H_1 the event that the defendant is guilty.

- D the event that the defendant's DNA matches DNA found at the crime scene.
- $P(D|H_1)$ the probability of seeing event D if the defendant is actually guilty.
- $P(H_1|D)$ the probability that the defendant is guilty assuming the DNA match (event D).
- $P(H_1)$ the juror's personal estimate of the probability that the defendant is guilty, based on the evidence other than the DNA match. This could be based on his responses under questioning, or previously presented evidence.

Suppose, on the basis of other evidence, a juror decides that there is a 20% chance that the defendant is guilty. Suppose also that the forensic testimony was that the probability that a person chosen at random would have DNA that matched that at the crime scene is 1 in a million.

The event D can occur in two ways. (1) The defendant is guilty (with prior probability $P(H_1) = 0.2$), and thus his DNA is present with probability 1, i.e., $P(D|H_1) = 1$. (2) He is innocent (with prior probability $P(H_o) = 0.8$), and he is unlucky enough to be one of the 1 in a million matching people, i.e., $P(D|H_o) = 1/(10^6 - 1) \simeq 10^{-6}$. Therefore,

$$P(D) = P(H_1) P(D|H_1) + P(H_o) P(D|H_o)$$
$$= 0.2 \times 1.0 + 0.8 \times 10^{-6} = 0.2000008.$$

Thus, the juror could coherently revise his opinion to take into account the DNA evidence based on Bayes' rule:

$$P(H_1|D) = \frac{0.2 \times 1.0}{0.2000008} = 0.999996.$$

This posterior probability is thought to be enough for the guilty verdict. In fact, even if the probability of guilty without DNA were $P(H_1) = 0.001$, the probability of guilty with DNA would be $P(H_1|D) = 0.999$. It seems that as long as a DNA match exists there is enough evidence for a guilty verdict.

THE PARADOX OF DNA IN THE JUDICIAL SYSTEM

However, DNA use in the courtroom is very controversial, as we discussed earlier (see DNA Paradox in Justice System). The challenging issues include possible fraud in the DNA testing lab, that crime scene samples were mixed up or contaminated with the suspect's comparison sample, that the suspect's DNA was intentionally planted at the scene, that there was some innocent reason for the suspect's DNA to be at the scene, that the suspect has a twin who might have left the DNA, that one of the above occurred and other samples that would have shown a different DNA pattern were either not tested or not reported to the jury, that the forensic testimony as to the odds was incorrect, and so forth (wikipedia.org). Even more complicated, there is the so-called multiplicity issue (see later in this chapter).

The two most amazing and long-lasting paradoxes in plausible reasoning might be the Paradox of Lewis Carroll's Urn and the Two-envelope Problem. These two paradoxes are as simple as they appear to be, yet as challenging as they are (I marked them with an asterisk).

2.3.2 *Paradox of Lewis Carroll's Urn**

Lewis Carroll (1832—1898) is the pen name of an English mathematician, professor of Oxford University and writer, C. L. Dodgson, whose literary legacy is well known. Among his literary works, the most widely known is *Alice in Wonderland*, which has been published many times and translated into many languages.

Lewis Carroll's Urn is a well-known puzzle in probability theory. Suppose that an urn contains two balls, each of which is either black (B) or white (W) with equal probability; thus, in the obvious notation, $P(BB) = P(BW) = P(WB) = P(WW) = 1/4$. Paradoxically and astoundingly, Carroll "proved" that there must be one white and one black ball in the urn. His method seems very straightforward and elegant as described below (Vakhania, 2009).

(1) We know that if a bag contained 3 balls, 2 being black and one white,

the chance of drawing a black ball would be 2/3; and any other state of things would not afford this chance.

(2) Now add a black ball to the urn, so that $P(BBB) = P(BWB) = P(WBB) = P(WWB) = 1/4$.

(3) Next we pick a ball at random; the chance that the ball is black can be calculated by the total probability:

$$1\left(\frac{1}{4}\right) + \frac{2}{3}\left(\frac{1}{4}\right) + \frac{2}{3}\left(\frac{1}{4}\right) + \frac{1}{3}\left(\frac{1}{4}\right) = \frac{2}{3}.$$

Given this probability, 2/3, and (1), there must now be two black balls and one white ball, which is to say that originally there was one of each color in the urn.

What do you think of the above "proof" that an urn cannot contain two balls of the same color? There seems nothing wrong with it. But paradoxically, we knew in the very beginning of the possibility of BB and WW.

This paradox has been studied widely for a long time. Most recently, (Vakhania, 2009) generalized the puzzle (Lewis Carroll's probability problem No. 72) for any number of black and white balls in the urn. Vakhania remarked: "The author of this note thinks that essentially there is no fallacy in the Lewis Carroll's proof. We could say it more definitely if we had not felt the pressure of the opposite opinion on this matter in all published papers concerning this problem." However, I believe the hidden fallacy becomes obvious if we rephrase the problem in the following way:

(1) Randomly select a white or black ball with $P(B) = 0.5$, and put it into the urn;

(2) Randomly select a white or black ball with $P(B) = 0.5$, and place it in the urn.

(3) Select black ball with $P(B) = 1$, and place it into the same urn.

Determine the colors of the first two balls placed in the urn.

In the probabilistic sense, i.e., under repeated experiment, there are not three fixed balls in the urn. The composition of three balls changes from time to time. Therefore, the total probability calculated above does not apply to any fixed-balls situation (e.g., three fixed balls: WBB). Consequently, Carroll's first statement above: " *if a bag contained 3 balls, 2 being black and one white, the chance of drawing a black ball would be 2/3; and any other state of things would not afford this chance*" is not applicable.

Another way to look at the problem is this: Suppose you don't know how I get two balls to the urn. Nevertheless, you add a black ball into the urn each time when I place the two balls, and then you randomly draw a ball. After repeating this many, many times you find that with 2/3 chance you draw a black ball. Following Carroll's logic, you would conclude that I always place one black and one white ball in the urn. But then you would almost certainly be completely wrong because I could actually place/replace the first ball with probability $P(B) = r$ and $P(W) = 1 - r$ and the second ball with $P(B) = 1 - r$ and $P(W) = r$. Here r can be any number between 0 and 1, inclusively. Even more strangely, I can use a different r each time I replace the two balls. Therefore, the colors of the two balls in the urn vary from time to time.

2.3.3 Two-Envelope Paradox*

The envelope paradox dates back to at least 1953, when Belgian mathematician Maurice Kraitchik proposed this puzzle (wikipedia.org):

Two people, equally rich, meet to compare their wallets' contents. Each is ignorant of the contents of the two wallets. The game is this: Whoever has the least money receives the contents of the wallet of the other (if the amounts are equal, nothing happens). One of the two men reasons: "Suppose that I have the amount A in my wallet. That's the maximum that I could lose. If I win (probability 0.5), the amount that I gain will be more than A. Therefore the game is favorable to me." The other man reasons in exactly the same way. However, by symmetry, the game is fair. Where is the mistake in the reasoning of each man?

In 1989, Barry Nalebuff presented the paradox in the two-envelope form as follows: The player is presented with two indistinguishable envelopes. One envelope contains twice as much as the other. The player chooses an envelope to open, receiving the contents as his reward. The question is, after the player picks an envelope without opening it, do you think he should switch to pick the other envelope if he is allowed?

Intuitively, you may say switching or not will make no difference because no information has changed. However, the following reasoning leads to a paradox.

(I) Let A and B be the random variables describing the amounts in the left and right envelopes, respectively, where $B = A/2$ or $B = 2A$ with equal probability.

(II) If the player holding the left envelope makes the decision to swap envelopes, he will take the value $2A$ with probability $1/2$ and the value $A/2$ with probability $1/2$.

(III) The "expected value" the player has by switching is $\frac{1}{2}(2A) + \frac{1}{2}\frac{A}{2} = 1.25A > A$. So a rational player shall select the right envelope expecting a greater reward than from the left envelope. However, since the envelopes are indistinguishable, there is no reason to prefer one to the other.

THE TWO-ENVELOPE PARADOX

There are various "solutions" to the paradox, but none of them really satisfy me. Therefore, I am going to analyze the paradox using examples from simple to complex. First, we clarify any confusion about the paradox by constructing some examples.

Example 1

(1) Generate randomly $x = 1$ or 2 with probability 0.5.
(2) Randomly put x or $y = 2x$ into envelope A with probability 0.5, and put whatever is left into envelope B. We also denote by A and B the values in the two envelopes, respectively.
(3) The sample space consists of four possible outcomes in the envelopes with equal probabilities: $\{(A, B)\} = \{(1, 2), (2, 1), (2, 4), (4, 2)\}$.
(4) Therefore, when $A = 2$, there is an equal probability (0.5) for $B = 1 = A/2$ and $B = 4 = 2A$.
(5) However, if $A = 1$ or 4, then $B = 2$ with probability 1.

Example 2

(1) Generate randomly $x = 1$ or 2 with probability 0.5.
(2) Randomly put x into envelope A or B with probability 0.5.
(3) Randomly put $x/2$ or $2x$ with probability 0.5 into the empty envelope (A or B).
(4) The sample space is:
$\{(A, B)\} = \{(1, 1/2), (1/2, 1), (1, 2), (2, 1), (2, 1), (1, 2), (2, 4), (4, 2)\}$.
(5) Therefore, when $A = 2$, $B = 1 = A/2$ with (conditional) probability 2/3 and $B = 4 = 2A$ with 1/3 probability.

(6) When $A = 1$, $B = 1/2 = A/2$ with probability $1/3$ and $B = 2 = 2A$ with probability $2/3$.

From Examples 1 and 2, we can see that the second statement is the paradox: "If the player holding the left envelope makes the decision to swap the envelop, he will take the value $2A$ with probability $1/2$ and the value $A/2$ with probability $1/2$." That which appears to be straightforward is really not necessarily true. Thus, the claim (III) is invalid. Conceptually, A is a random variable not a constant, thus saying "the expected value is $1.25A$" is nonsense.

Example 3

(1) Generate randomly x from a proper distribution $f_X(x)$, $x > 0$.
(2) Randomly put x or $y = 2x$ into envelope A with probability 0.5, and whatever is left into B.
(3) Thus, we have $X \sim f_X(x)$, $Y \sim f_Y(Y = x) = f_X(2x)$, $A \sim \frac{1}{2}(f_X(x) + f_X(2x))$, and $B \sim \frac{1}{2}(f_X(x) + f_X(2x))$.
(4) Therefore, for a typical x, $P(B = x/2|A = x) \neq P(B = 2x|A = x) \neq 0.5$.

Therefore, the claim (III) is invalid. Readers can analyze this example for an exponential distribution: $f_X(x) = \lambda \exp(-\lambda x)$.

Example 4

We randomly assign 2^n and 2^{n+1} to envelopes A and B, after randomly choosing an integer n from $\{..., -2, -1, 0, 1, 2, 3, ...\}$. The sample space is: $\{(2^{-(n+1)}, 2^{-n}), \cdots, (\frac{1}{8}, \frac{1}{4}), (\frac{1}{4}, \frac{1}{2}), (\frac{1}{4}, \frac{1}{8}), (\frac{1}{2}, 1), (\frac{1}{2}, \frac{1}{4}), (1, 2), (1, \frac{1}{2}), (2, 4), (2, 1), (4, 8), (4, 2), (8, 16), (8, 4), (16, 8), \cdots, (2^{n+1}, 2^n)\}$.

We can see that, for any k between $-n$ and n, inclusively, we have $P(B = 2^{k-1}|A = 2^k) = P(B = 2^{k+1}|A = 2^k) = 0.5$. But because of the first elementary events $(2^{-(n+1)}, 2^{-n})$, we know that $P(B = 2^{-n}|A = 2^{-(n+1)}) = 1$ and $P(B = 2^{-(n+2)}|A = 2^{-(n+1)}) = 0$. Hence, $P(B = 2^{-n}|A = 2^{-(n+1)}) \neq P(B = 2^{-(n+2)}|A = 2^{-(n+1)})$ (no matter how large n is). Similarly, $P(B = 2^n|A = 2^{n+1}) \neq P(B = 2^{n+2}|A = 2^{n+1})$. Therefore, the claims (II) and (III) are false. Furthermore, from the sample space it can be seen clearly that to switch the two envelopes is actually switching the two (equal) expected values. We will not gain anything by switching.

In summary, we have seen how it is not possible to make two envelopes with properties (I) and (II).

The last set of paradoxes in this chapter will be used to illustrate that properties in deterministic mathematics may not be applied to probability directly. There are distinctions between mathematical or formal reasoning and plausible reasoning.

2.3.4 *Paradoxes of Random Variables*

A function and a random variable have many similarities and differences. A random variable is a function of a random event. If we attempt to infer a property of a random variable as one might do for a function, we can be easily be trapped.

Paradox of Distributional Similarity

If two variables X and Y so that $X = Y$ and Z is another variable, then $XZ = YZ$. Now if random variables X and Y have the same distribution, denoted by $X \stackrel{d}{=} Y$, then we would expect $XZ \stackrel{d}{=} YZ$ for any random variable Z. Surprisingly, this is not true in general. Stoyanov (1997, p. 31) gives the following example: Suppose random variable X and Y have symmetric distributions and $Y = -X$. It is obvious that $X \stackrel{d}{=} Y$. Now let $Z = Y$ so that $XZ = -X^2$ and $YZ = X^2$. that is, $Z = -X$. The equality $XZ \stackrel{d}{=} YZ$ is impossible because all values of XZ are nonpositive while those of YZ are nonnegative.

If X and Y are identical functions, then $X + Y$ determines uniquely the function X, i.e., $X = (X + Y)/2$. Similarly, if X and Y are independent, identically distributed random variables, we would expect that the distribution of $X+Y$ determines uniquely the common distribution of X and Y; but, paradoxically, this is not always the case. An example is given by Székely (1986, p. 125), by means of characteristic functions.

Furthermore, there exist random variables X, Y, and Z such that the probability distribution of $X+Y$ is equal to that of $X+Z$, but the distributions of Y and Z are not the same.

A Paradox of Expectation

Székely (1986, p.193) presents an interesting paradox about the expectation of the sum of random variables: "It is wellknown that if X and Y are random variables and the expectation $E(X + Y)$ exists, then $E(X + Y)$ can be determined without knowing the joint distribution of X and Y. Surprisingly, this is not true for three variables. In other words, if X, Y, and Z are arbitrary random variables for which $E(X + Y + Z)$ exists, then this expectation cannot always be determined knowing only the individual distributions of X, Y and Z. Furthermore, since $E(X + Y + Z) = E((X + Y) + Z)$ and $E(X + Y + Z + W) = E((X + Y) + (Z + W))$, the expectations of sums of three and four variables are uniquely determined by the two-dimensional distributions. It is not known, however, whether there are similar conclusions for more than four random variables."

2.4 Chapter Review

We discussed the apparent difference in frequentist and Bayesian notions of probability, revealed the hidden assumption (sample space) of the Bayesian notion of probability and unified the conceptions of probability. We use (1) the Boy and Girl Paradox to show how the sample space will depend on the question at hand, (2) Coupon Collector's Paradox to study the sequential sampling issue, and (3) Bertrand's Paradox to illustrate how confusing the terms "random" or "equal probability" could be in probability.

The Monty Hall Dilemma has provoked widespread discussion. It has fooled many PhDs in statistics. It reveals the importance of hidden assumptions and how they might differ between frequentist and Bayesian perspectives in terms of decision making. It also signifies how the host's intention may or may not affect a player's calculation of probability, which will be further discussed in the next chapter.

The Tennis Game Paradox is of special interest example of probability. It shows counterintuitively a situation where you have more chances of winning when competing with a group of higher-ranked individuals than lower-ranked individuals.

Ruining time problems are commonly encountered in daily life and engineering. The Paradox of Ruining Time is used to show how counterintuitive a ruining time can be.

The Paradox of Nontransitive Dice is used to illustrate that the properties in (deterministic) mathematics cannot always be valid when the probability is involved. Similar issues are addressed in Paradoxes of Random Variables.

Independence of events is a common phrase in probability, but there are several different concepts of independence. We clarify the differences in the Paradox of Independence.

The most obvious concept is often the most fundamental and most difficult to explain. Randomness is an archetypal example. Probability is a study of random phenomena, but what is randomness? Randomness is defined by complexity or incompressibility in mathematics and computer science. It is defined by disorder and measured by entropy in information science. Unfortunately, randomness is not operationally well defined in probability and statistics so we cannot determine with certainty that an event or a sequence of events is random or not.

Quantum probability challenges the notion of a deterministic world, i.e., anything is determined by its cause and the cause of the cause, and so on. Quantum mechanics thus made the first radical shift away from the classical concept that everything in the universe is perfectly determined.

The conceptual discussion of probability serves as a basis for plausible reasoning since we are often dealing uncertain events. Having said that, it is natural to study the formal logic (mathematical logic) before plausible reasoning since the former is a part of plausible reasoning.

We introduced conceptually induction, deduction, and, causal inference. We discussed the basics of propositional calculus and how to use the elementary laws in proportional calculus to convert a formal reasoning problem into an algebra or arithmetic problem.

Reasoning is a process to search out the truth, while the law of the excluded the middle asserts one and only one statement is correct between a statement and its opposite statement. However, the Paradox of Truth shows that two contradictory conclusions or statements can both be correct since the statement itself affects the outcome.

Mathematical Induction is a widely used tool for problems that can be indexed by integers. It is so powerful, is so widely used, and requires so little effort to learn that I introduced it with a simple example.

In scientific reasoning, we are often attracted by things that are appealing and loud out there, and intentionally or unintentionally neglect other things which otherwise are informative. The Three-Hat Puzzle illustrates how "silence" can be quite informative.

For plausible reasoning, we introduce Bayesian reasoning using Bayes' rule with an example of DNA evidence used in a courtroom. The numerical example gives you a sense of the reasoning tool.

The two most amazing and long-lived paradoxes in plausible reasoning might be the Paradox of Lewis Carroll's Urn and the two-envelope problem. These two paradoxes are as simple as they appear to be, yet as challenging as they are (I have marked them with asterisk). They are unsettling and intriguing.

The Paradox of Lewis Carroll's Urn is a mixture of formal reasoning and plausible reasoning. The paradox is so simple and straightforward that you can't seem to find any error in the reasoning, but the result is so ridiculous that you know something is wrong. Even a recent publication of the *Bulletin of the Georgian National Academic of Sciences* deems this an unsolved puzzle. We pointed out in a fairly simple way where the fallacy of Lewis' reasoning is.

The Two-Envelope Problem is one of the most widely studied paradoxes. So far no claimed solutions are satisfactory. We analyze four different examples that show how it is impossible to meet the conditions claimed in the game – I cannot see how it possible to construct the game.

The last set of paradoxes illustrates distinctions between mathematical or formal reasoning and plausible reasoning.

2.5 Exercises

2.1 When you decide whether or not to buy a lottery ticket, what are you thinking in terms of probability? How do you determine the probability or chance of winning?

2.2 When you make an investment such as buying a stock or allocating your retirement funds, what is your decision process and how and where did your data or information come from? How does the concept of probability play a role in your decision and what is the conceptual probability in this case?

2.3 When you said (I assume you have said) I was lucky to have achieved something or avoided some accident, what did you mean? Did you mean you were supposed to have a high probability of having it but luckily or unluckily it didn't happen? Where does this probability come from?

2.4 What is a sample space? Are the probabilities for the outcomes in a sample space equal? What is an outcome space? Are the probabilities for outcomes in the outcome space equal?

2.5 We often find that the first 80% of work needs 20% of the effort and the remaining 20% of work needs 80% of the effort. Can you use the Coupon Collector's Problem to illustrate why?

2.6 For the Monty Hall Dilemma, prove the following:

(1) If the host is determined to show you a goat, and picks one at random out of a choice of two, the probability p would be $2/3$.

(2) If the host is determined to show you a goat, with a choice of two goats (Bill and Nan, say), and he shows you Bill with probability n, then given that you are shown Bill, the probability that you will win by switching doors is $p = 1/(1 + n)$.

(3) If the host opens a door chosen at random, irrespective of what lies behind it, then $p = 1/2$.

If the host opens a door with a goat without telling you how, would you switch or not? You can use the Bayesian approach, construct the prior for the host's intention, and calculate the posterior probability for your decision-making. If the host randomly selects a door to open and, if it is a goat, offers you the option to switch, would you want to switch?

2.7 In Exercise 2.6, the different values of probability clearly indicate that the host's intention will affect one's decision. Would the host's intention affect your decision? Do you think the host's intention for the past or future game will affect your decision in the current game? Why or why not?

2.8 The Tennis Game Paradox is an example that shows that it is easy to defeat a group with stronger individual skills than a group with weaker individual skills. Of course, the level of individual skill varies. Can you give other examples in which a group has a stronger aggregated skill but lower individual skills (Hint: swarm intelligence)?

2.9 For scientific research, first-hand data often play an important role. Make your standard six-sided dice into Efron's dice and two-way round-defeat dice. Verify the round-defeat properties by repeating the experiment: Throw the dice many times.

2.10 In the Paradox of Ruining Time, the derivation of the ruining time mn is straightforward: We know the total fortune of each player over time can be viewed as a Brownian motion, and from the previous section we know the average time (number of rounds) required to hit the boundaries (either player wins) is m^2 and n^2 with probability $n/(m+n)$ and $m/(m+n)$, respectively. Therefore, the expected ruining time is the weighted average: $m^2 \frac{n}{m+n} + n^2 \frac{m}{m+n} = mn$.

To obtain the wining probability $P(m)$, the key to obtaining this probability is to construct a recursive formulation. The player has probability p of winning \$1, and then would have a total fortune of $m+1$ dollars after the first round. At same time he has probability q of losing \$1 and instead having a total fortune of $m-1$ dollars after the first round. Then the game continues as before with the same winning probability function but different "initial values," $m+1$ and $m-1$, respectively. Therefore, we obtain the equation:

$$P(m) = p \cdot P(m+1) + q \cdot P(m-1).$$

Using the variable substitutions $x_m = P(m) - P(m-1)$, the above recursion becomes a geometric series: $x_m = \left(\frac{q}{p}\right) x_{m-1}$. By using the boundary conditions of these ordinary difference equations, which are $P(0) = 0$ and $P(m+n) = 1$, we obtain the solution $P(m)$.

An interesting question is: Would the player be better off playing the game with some dollar amount y instead of 1 dollar bets? What is the optimal y?

2.11 Prove the *contrapositive law*: $(A \longrightarrow B) \longleftrightarrow (\neg B \longrightarrow \neg A)$, the *transitivity of implication*: $(A \longrightarrow B) \wedge (B \longrightarrow C)) \longrightarrow (A \longrightarrow C)$, *proof by cases*: $((A \longrightarrow B) \wedge (\neg A \longrightarrow B)) \longrightarrow B$, and *De Morgan's rules*: $\neg(A \vee B) \longleftrightarrow (\neg A \wedge \neg B)$ and $\neg(A \wedge B) \longleftrightarrow (\neg A \vee \neg B)$.

2.12 A short reading assignment:

Fermat's last theorem, discussed in Chapter 1, can be stated in terms of the predicate calculus (x, y, z, and n are positive integers):

$$\forall n \forall x \forall y \forall z (n > 2) \longrightarrow x^n + y^n \neq z^n.$$

In 1993, Gödel proved his incompleteness theorem, which states that, in any logical system complex enough to contain arithmetic, it will always be possible to find a true result that is not formally provable using predicate logic. This result was especially startling because the notions of truth and provability had usually been identified with each other.

A formula is said to be logically valid (or a tautology) if it is true for every interpretation. For example, the negation of a formula is characterized by the identities below:

$$\neg \forall x \; P(x) = \exists x \; \neg P(x),$$

which in layman's term can be stated as: "Not all x have property $P(x)$" is equivalent to "Some x do not have property $P(x)$."

2.13 We have described the concepts of independent events in Paradoxes of Independence. Do you believe that events matching this description exist in the real world? Why or why not?

2.14 Most everyone has used or heard the words *random* and *randomness*. How do you define these terms? Can you tell which of the following is a random sequence of letters and which is not: (1) "concept," (2) "cnocpet," and (3) "ceptcon?" Do you determine randomness by the meaning of the sequence of letters, the mechanics of generating the sequence, or something else?

2.15 What is your definition of truth? Can you tell the difference between the two definitions of truth: (1) everything that can be proven correct and (2) everything that cannot be proven incorrect (more precisely, everything that, it has been proven, cannot be proven incorrect)? Give an example that shows the difference between the two definitions.

2.16 Use Mathematical Induction to prove:

$$1^2 + 2^2 + 3^2 + \cdots + n^2 = \frac{1}{6}n(n+1)(2n+1), \text{ and}$$

$$1 \cdot 2 + 2 \cdot 3 + \cdots + n(n+1) = \frac{1}{3}n(n+1)(n+2).$$

2.17 List as many possible meanings of "silence" as you can.

2.18 We have given an example on how to use Bayesian reasoning and DNA evidence in a courtroom. Do you believe DNA evidence should be used in a courtroom? Elaborate your answer.

2.19 Explain (in your own words) the fallacy of Lewis Carroll's reasoning in his paradox about the urn.

2.20 We use four examples to illustrate that we cannot meet the conditions in the two-envelope problem. Try to see if you can describe a process that meets the two conditions:

(1) Let A and B be the random variables describing the amounts in the left and right envelopes, respectively, where $B = A/2$ or $B = 2A$ with equal probability.
(2) If the player holding the left envelope makes the decision to swap envelopes, he will take the value $2A$ with probability $1/2$ and the value $A/2$ with probability $1/2$.

2.21 Can you think of any applications from what you have learned from this chapter?

Chapter 3

Statistical Measures of Scientific Evidence

Quick Study Guide

This chapter is given to further advance probabilistic inference to statistical inference. However, this chapter is not to teach you hands-on statistical methodologies, nor are you expected to perform statistical analyses. Instead, we will provide in-depth discussions and critiques on fundamental statistical principles and different statistical paradigms that are derived from these principles. Each of the principles is appealingly intuitive but the surrounding controversies are extremely complex. We will use paradoxes in numerical forms to effectively explain the abstract concepts and complex issues. Even so, a great patience is required to fully benefit from this chapter.

We will describe the three statistical paradigms frequentism, Bayesianism, and likelihoodism, and the Decision Approach as well. After reviewing the concepts of statistical model, point estimate, confidence interval, p-value, type-I and type-II errors, and level of significance, we will provide a detailed discussion of five essential statistical principles and surrounding controversies: the conditionality principle, the sufficiency principle, the likelihood principle, the law of likelihood, and Bayes' law.

The discussion of controversies will be broadened to statistical analyses, including model fitting, data pooling, subgroup analyses, regression to the mean, and the issue of confounding. We put a large effort on the multiplicity issues due.

We will make an attempt to unify the statistical paradigms under the proposed principle of evidential totality and the new concept of causal space.

This chapter covers comprehensively and in depth most controversial issues in statistical inference. To make it effective, I will again borrow from the power of paradoxes.

3.1 Introduction to Statistical Methods

3.1.1 *Statistics versus Probability*

We have just discussed the concept of probability. In reality, we are more often dealing with statistics in quantifying scientific evidence. For frequentism, Probability and Statistics differ. Probability arises to quantify the relative frequencies of errors in a hypothetical long run; for example, probability describes what a sample of results might be when drawn from the entire population of results, and when the entire population is already known (Cox and Mayo in Mayo and Spanos, 2010). Statistics operates in the opposite direction. A statistical analysis starts with a known sample, and attempts to infer what the population might look like from the information in the sample, and knowing how the sample was created (Chang, 2008). For Bayesianism, probability enables one to quantify the "rational" degree of belief, providing confirmation, or credibility in hypotheses. While Frequentist and Bayesian approaches do calculate statistics and probability differently, a conceptual unification exists and is discussed under "Causal Space Theory" in the previous chapter.

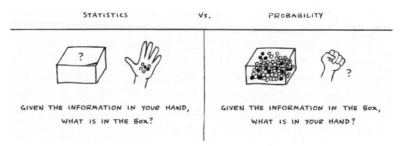

STATISTICS   Vs.   PROBABILITY

GIVEN THE INFORMATION IN YOUR HAND,   GIVEN THE INFORMATION IN THE BOX,
WHAT IS IN THE BOX?   WHAT IS IN YOUR HAND?

3.1.2 *Statistical Paradigms*

As pointed out by Cox and Mayo (Mayo and Spanos, 2010), statistical methods are used virtually in all areas of science, technology, public affairs, and private enterprise. The variety of applications makes any single unifying discussion difficult if not impossible. Objectivity in statistics, as in science more generally, is a matter of both aims and methods. Objective science aims to find out what is the case as regards aspects of the world, independently of our beliefs, biases, and interests; objective methods thus aim at critical control of inferences and hypotheses, constraining them in particular by evidence and checks of error.

Fisher (1935, p.39) said: "It should never be true, though it is still often said, that the conclusions are no more accurate than the data on which they are based. Statistical data are always erroneous, in greater or lesser degree. The study of inductive reasoning is the study of the embryology of knowledge, of the processes by means of which truth is extracted from its native ore in which it is fused with much error."

There are mainly three statistical paradigms in terms of philosophical and methodological differences: Frequentism, Bayesianism, and Likelihood-ism. Controversies among them have rumbled on for anywhere from 50 to 200 years. I will outline the key differences and issues through paradoxes and set the stage for elucidating a unifying principle.

Frequentist Paradigm

This paradigm makes use of the frequentist view of probability to characterize methods of analysis by means of their performance characteristics in a hypothetical sequence of repetitions. There are two main formulations used in the frequentist approach to summarizing evidence about a parameter of interest, ψ. The first is the provision of sets or intervals within which ψ is, in some sense, likely to lie (confidence intervals), and the other is the assessment of concordance and discordance with a specified value, θ (significance tests) (Cox and Mayo in Mayo and Spanos, 2010) .

The key difference between the frequentist approach and the other paradigms is its focus on the sampling distribution of the (test) statistic, that is, its distribution in hypothetical repetition. In Cox's view, the sampling distribution, when properly used and interpreted, is at the heart of the objectivity of frequentist methods.

Bayesian Paradigm

"Bayesian" refers to the 18th century mathematician and theologian Thomas Bayes (1702–1761), who provided the first mathematical treatment of a nontrivial problem of Bayesian inference. Nevertheless, it was the French mathematician Pierre-Simon Laplace (1749–1827) who pioneered and popularized what is now called Bayesian probability.

In the Bayesian approach, the parameter is modeled as a realized value of a random variable with a probability distribution called the prior distribution. Having observed y, inference proceeds by computing the conditional distribution, given $Y = y$, of the posterior distribution.

Likelihood Paradigm

The likelihood approach rests on a comparative appraisal of rival statistical hypotheses, H_o and H_a, according to the ratio of their likelihoods. The basic premise is what Hacking (1965) called the law of the likelihood: that the hypothesis with the higher likelihood has the higher evidential "support," and so is the more "plausible." (Cox and Mayo in Mayo and Spanos, 2010). Operationally, this approach is built on the ratio of likelihoods of two different parameter values. The ratio, depending only on a statistic, is regarded as a comparative summary of "what the data convey."

3.1.3   *Statistical Models*

A statistical model, $M = \{p(x;\theta) : x \in \chi, \theta \in \Theta\}$, is a set of possible probability distributions, $p(x;\theta)$, for the observed data x from sample space χ and the parameter θ belonging to the parameter space Θ. Here, $\in$ means "belongs to" or "is from." In some cases there may be other information, such as a prior distribution for the parameter, $\pi(\theta)$. Therefore, statistical evidence from the current experiment can be denoted by (x, M).

The uncertainties of the inferences (statements or conclusions) are an essential part of the inference. The statements and their uncertainties are both based on statistical evidence and together form the evidential meaning of the statistical evidence.

Example: Negative Binomial Distribution

Consider a Bernoulli trial with an unknown probability of success, p. Suppose the trial will continue until a predetermined number of successes, r, is reached. Denote the total number of trials needed by $r + X$, where random variable X is the number of failures, which has a negative binomial distribution. The minimum variance unbiased estimator of p is given in terms of the observed number of failures, x (Johnson, Kotz, and Kemp, 1993, p.215):

$$\hat{p} = \frac{r-1}{r+x+1}.$$

Paradox of Confidence Intervals

We know that a confidence interval (CI) is an interval estimate of a population parameter and is used to indicate the precision of an estimate. A confidence is constructed from observations using a valid statistical model, thus it varies from sample to sample. A CI frequently covers the parameter of interest, if the experiment is repeated. How frequently the observed interval contains the parameter is determined by the confidence level or confidence coefficient. For instance, a 95% CI is expected to cover the true value of the parameter 95% times if the experiment is repeated sufficiently many times.

The calculation of a confidence interval usually involves point estimate of a parameter of interest and the variability of the sample estimate. Surprisingly, it has been known since the 1960s that from a single observation x from a normal distribution with unknown mean μ and unknown standard deviation, it is possible to create a confidence interval (CI) for μ with finite length. This remarkable result seems to completely contradict the standard statistical intuition that at least two observations are necessary in order to have some idea about variability.

For any $\alpha \in (0,1)$, there exists a finite $100(1 - \alpha)\%$ CI for the mean of any unimodal distribution, of the form

$$x \pm \eta |x|,$$

where the constant η is solely a function of the size α, and satisfies

$$P\left(X - \eta |X| \leq \mu \leq X + \eta |X|\right) \geq 1 - \alpha.$$

Furthermore, the value of η decreases with increasingly strong assumptions about the underlying distribution (unimodal, unimodal-symmetric, or normal). Blachman and Machol (1987) derived an approximation to η for a normal $X \sim N\left(\mu, \sigma^2\right)$, i.e., $\eta \approx \frac{2\phi(1)}{\alpha}$, where ϕ is the standard normal density. Wall, Boen, and Tweedie (2001) show that CIs with a sample of size $n = 2$ from a normal distribution can yield better results in some cases than the usual Student's t interval.

This example shows that x itself tells us something about the variability as well as the mean.

3.1.4 *Paradox of Classification*

Suppose we need to classify a subject to a group with parameter $\theta = \theta_1 = 1$ (the healthy) or the group with $\theta = \theta_2 = -1$ (the unwell). We have four measures of θ from the subject: $x = \{x_1, x_2, x_3, x_4\} = \{-3, -2, 1.5, 3.5\}$. We know that the measurements involve some random errors, and define the error as

$$d\left(\theta | x\right) = \frac{1}{4} \sum_{i=1}^{4} |x_i - \theta|.$$

For the given data x, $d\left(\theta | x\right)$ is constant when $-2 \leq \theta \leq 1$. Therefore, if we consider $d\left(\theta | x\right)$ as the statistical evidence against the classification of a subject to a group based on the given data x, we can either assign the subject to θ_1 (healthy) or θ_2 (unwell) group. Paradoxically, if we define the statistical error as

$$d\left(\theta|x\right) = \frac{1}{4}\sqrt{\sum_{i=1}^{4}\left(x_i - \theta\right)^2},$$

then $d\left(\theta_1|x\right) = 31.5/4$ and $d\left(\theta_2|x\right) = 28.5/4$. Therefore, we would assign the subject to the unwell group.

From this example we can see that the measure of scientific evidence can be subjective and different measures can lead to different conclusions.

3.1.5 *Hypothesis Testing*

Those who don't have trainings in statistics may simply believe a person who made a prediction with 100% accuracy more than a person who made a prediction with a lesser (e.g., 99%) accuracy. But a statistician may be a little more sophisticated on this. For instance, a person predicts weather (rain or not) by flipping a fair coin. By chance he may predict the raining day correctly in his first prediction; thus, his accuracy is 1 out of 1, i.e. 100% accuracy. On the other hand, a meteorologist with years of training and research in the discipline may only predict the weather 800 times correctly out of 1000, i.e., 80% accuracy. We, of course, believe the latter more than the coin flipper. In other words, we want to avoid drawing conclusion based on a random chance. Thus, the notion of hypothesis testing is to control the probability of making false positive claim. There are two common forms for hypothesis testing.

(1) Fisher's hypothesis testing (domain test):

$$H_o : \theta \in \Theta_o \text{ or } H_a : \theta \in \Theta_a, \tag{3.1}$$

where Θ_o and Θ_a are the parameter spaces for the null and alternative hypotheses, respectively. They compose the whole parameter space or the set of all possible values for parameter θ. Θ_a is often taken to be the negation of Θ_o.

For example, in drug development, θ can usually represent the treatment effect of a new drug, $\Theta_o = (-\infty, 0]$ and $\Theta_a = (0, \infty)$. Thus, H_o means no or a negative treatment effect and H_a becomes a positive treatment effect.

(2) Neyman—Pearson hypothesis testing (point-to-point test):

$$H_o : \theta = \theta_0 \text{ or } H_a : \theta = \theta_a. \tag{3.2}$$

For convenience, we call (3.1) a domain-test, and (3.2) a point-to-point test. The probability of erroneously rejecting H_o when H_o is true is called the false positive rate or type I error rate. Under hypothesis testing in the frequentist paradigm, rejection of H_o has to ensure control the type I error rate at a nominal value, α, called the size of the test or level of significance. Similarly, the probability of erroneously rejecting H_a when H_a is true is called the type II error rate, which is also controlled at a predetermined level, β. The probability of rejecting H_o is the power of the hypothesis test, which is dependent on the particular value of θ. The power is numerically equal to $1 - \beta$ when H_a is true and equal to α when H_o is true.

For hypothesis testing (3.1), we can measure the strength of the evidence against H_o using the p-value, which is defined as the (least upper bound of the) probability of getting data the same as or more extreme than the observed data when the null hypothesis H_o is true. The p-value will be compared with a nominal threshold α (e.g., 0.05) to determine if the null hypothesis should be rejected.

When the null hypothesis is rejected, the p-value is the probability of the type I error (or type I error rate). It is unfortunate that many people refer to the size α as the type I error rate, causing unnecessary confusion.

Example: Hypothesis Testing

An insurance company is reviewing its current policy rates. They are concerned that the true mean claim amount, μ, is actually higher than the originally assumed average claim amount, $\mu = \$1800$. It is critical for company policy-making to find out if this concern is real. For this purpose, a random sample of 40 claims is conducted, and there is calculated a sample mean of $\bar{x} = \$1950$. Assume that the standard deviation of claims is $\sigma = \$500$, and set $\alpha = 0.05$. The hypothesis test can be stated as $H_o : \mu \leq 1800$ versus $H_a : \mu > 1800$. The test statistic is defined as

$$z = (\bar{x} - \mu) / (\sigma / \sqrt{n}).$$

The p-value is the maximum of the probability that the value of the random variable Z is equal to or larger than the observed value z when H_o is true. The maximum probability occurs when $\mu = 1800$. Thus, $z = 1.879$ and the corresponding p-value can be obtained from the standard normal distribution, i.e., $p = 0.03 < \alpha = 0.05$. Therefore, we reject H_o and conclude that a modification of the insurance policy may be necessary.

For the weather prediction problem, if someone can predict correctly for five days in a row, then we may believe he has some special knowledge or trainings in this area, because $0.5^5 = 0.031\,25 < \alpha = 0.05$.

Some Remarks

- The choice of the level of significance α is somewhat arbitrary; it is small due to the serious concern about the impact of the potential type I error
- The p-value is only dependent on the tail distribution of the statistic under H_o, but is independent of H_a. Hence, many sets of observations can lead to the same conclusion (reject H_o or not) as long as the percentage of their "tails" are the same.
- Failure to reject H_o does not mean H_a is true.
- When the sample size is large enough, no matter how small the difference of the parameter is, statistical significance can be achieved.

More importantly, the p-value is the conditional probability given H_o. The p-value is not the probability that H_o or H_a is true. It is usually true that $P(\theta \in \Theta_o) + P(\theta \in \Theta_a) \equiv 1$, meaning that the sum of the probabilities that H_o is true and that H_a is true is unity. However, in general $P(x|\Theta_o) + P(x|\Theta_a) \neq 1$. In other words, the probability of observing the data when the null is true and the probability of observing the same data when the alternative is true does not add up to unity. We should not be confused by the apparent similarity between $P(\theta \in \Theta_o) + P(\theta \in \Theta_a)$ and $P(x|\Theta_o) + P(x|\Theta_a)$.

The problem raised from the p-value being independent of the distribution of the test statistic under H_a can be illustrated in the following paradox. In the paradox, the upper tail distribution of the test statistic $f(x|H_o)$ is larger than that of $f(x|H_a)$, but the mean μ_x is smaller when the given H_o is true than that when H_a is true.

Paradox of Hypothesis Testing

A experimenter wrote down a number on a paper, but forgot whether is was a sample from the standard normal distribution $N(\theta, 1)$ with mean $\theta = 0$ or from the following the triangle distribution centered at mean $\theta = 0.5$.

$$x \sim f(x; \theta) = \begin{cases} 1 - |x - \theta| & \text{if } -0.5 \leq x \leq 1.5 \\ 0 & \text{otherwise.} \end{cases}$$

So he decided to use the following hypothesis test, $H_o : \theta = 0$ and $H_a : \theta = 0.5$.

The test statistic T is simply the sample value x_1, i.e., $T = x_1$. The question is: Should the distribution of the test statistic be a mixture of the $N(0, 1)$ and $f(x; \theta)$ distributions, or should two different distributions be used when we make an inference about θ (0, or 0.5)? Under the notion of repeated experiments, the T will have a mixture of two normal distributions. However, conditioning on H_o, (the subset of) the test statistic T is from $N(0, 1)$.

Based on the definition, the p-value is calculated under H_o, i.e., $N(0,1)$. Suppose we have observed $T = 1.96$, thus the p-value is 0.025; we reject H_0 at a one-sided $\alpha = 0.025$ and conclude that $\theta = 0.5$. However, the sample is from $N(0,1)$ because the maximum value of T is 1.5 if the sample is from $f(x; \theta = 0.5)$. Even though the type I error rate is controlled at the α level, we always make a mistake when we reject H_o using the critical value 1.96.

It is totally surprising that a small p-value does not necessarily indicate the evidence is against H_o or consistent with H_a. Instead the small p-value, in this case, indicates H_o is true and H_a is wrong. Making the significance level α even smaller (e.g., 0.001) would not change the situation. The likelihood ratio for H_a against H_o is zero, which implies H_a is impossible!

The difference in the distribution $f(x; \theta)$ under H_o and H_a in the Paradox of Hypothesis Testing may be not common, but $X \sim N(0,1)$ under H_o and $X \sim N(0.1, 1/10)$ under H_a can be common in practice.

Another simple but controversial example would be: Suppose we want to test the hypothesis $H_o : \theta = -1$ versus $H_a : \theta = 1$, knowing that the random variables are $X \sim N(\theta, 0.25)$, with θ either being 1 or -1. Presumably, the level of significance is chosen to be $\alpha = 0.05$. Suppose a single observation $x = 0$ was collected. Thus, the corresponding p-value is $0.0228 < 0.05$, and H_o is rejected. The question is: Is this $(x = 0)$ really strong evidence against H_o and in favor of $H_a : \theta = 1$?

3.1.6 *Spring Water Paradox*

If controlling the type I error rate is a critical goal for an experiment, the following is a paradox. An experiment is to be conducted to test the hypothesis that spring water is effective as compared to a placebo in a certain diseased population. To carry out the trial with a minimal cost, we will need a coin and two patients with the disease. To control type I error, the coin is used, i.e., **0** patients are needed. To have an unbiased point estimate, a sample-size of **2** patients is required (for a higher precision, more patients are needed). The trial is carried out as follows:

(1) Randomize the two patients: One takes the placebo and the other drinks spring water. The unbiased estimate of the treatment difference is given by $y - x$, where x and y are the clinical responses from the placebo and the spring water, respectively.

(2) If $y - x > 0$, we will proceed with the hypothesis test $H_o : \mu_y - \mu_x \leq 0$ at a significance level (one-sided) of $\alpha = 0.05$. To perform the hypothesis test, we flip the coin 100 times. If heads appears at least 95 times, the

efficacy of spring water will be proclaimed.

THE SPRING WATER PARADOX

Suppose the spring water is very safe. This is a cost-effective approach (two patients and a coin), even though there is only approximately a power of 5% and approximately a 2.5% probability of making the efficacy claim.

What if all pharmaceutical companies test such water-like compounds, over and over again, in their drug development? There will be many ineffective drugs in the market. In fact, this scenario could arise for two reasons: (1) A small α makes it difficult to show statistical significance for compounds that have small-to-moderate effects, so testing water-like compounds is more cost effective. (2) Many companies screen the same compound libraries without multiple-testing adjustments. Therefore, a smaller α (more stringent type I error control) could lead to more ineffective compounds flowing into the drug development pipeline and to the market.

3.2   Statistical Principles

Like any other science, Statistics has its own principles (axioms) that do not require proof but have to be followed. Different principles will lead to different statistical paradigms, which in turn could provide different answers to the same question. This controversy has raised the issue about what the right or best approach is to the problem at hand, or in general which paradigm is the best in which circumstances.

3.2.1   *Conditionality Principle*

The Conditionality Principle can be stated as: If m experiments on the parameter θ are available with equal probability of being selected, the resulting inference on θ should only depend on the selected experiment.

Paradox of Confidence

Suppose two observations, X_1 and X_2, are taken from the symmetric probability mass function:

$$P_\theta (X = \theta - 1) = P_\theta (X = \theta + 1).$$

To estimate the parameter θ, we can, based on the conditionality principle, propose the following estimator:

$$\delta (X) = \begin{cases} X_1 - 1, \text{ if } X_1 = X_2 \\ \frac{X_1 + X_2}{2}, \text{ if } X_1 \neq X_2 \end{cases}$$

To a frequentist, without conditionality, this estimate has a confidence level of 75% for all θ, that is, $P (\delta (X) = \theta) = 0.75$. However, the conditionalist would report confidence of 100% if the observed data in hand are different, or 50% if the observations coincide.

The question is: Does it make sense to report the preexperimental accuracy, which is known to be misleading after observing the data?

Paradox of Instrument Precision

Suppose two instruments are available with an equal chance to an experimenter, one with 0.001 precision and the other with 0.1 precision. Should the experimenter report the results, based on the instrument actually used in the experiment, or should the report be based on the mixed precision of both available instruments (Cox, 2006)? If the answer is the latter, then in reality all instruments in the world can be considered available to the experimenter, with different probabilities. How should he determine the precision of his experiment?

Frequentism's Violation of Conditionality Principle

The frequentist hypothesis testing paradigm violates the conditionality principle as further illustrated in the following. It is the common view that conditionality is adapted by frequentism, Bayesianism, and likelihoodism. However, that is not necessarily true. Let's take a close look at the Paradox of Hypothesis Testing. If H_o is true; x_1 comes from $N(0,1)$, and we are making an inference based on the experiment that is actually performed. However, if H_a is true; x_1 is from $f(x; 0.5)$, and based on the conditionality principle, the inference about θ should not be dependent on $N(0,1)$ at all, because the corresponding experiment was not performed. But we make an inference on θ based on a p-value that is calculated through the distribution of the statistic asserted by H_o, even though H_a is true. Even more ironically, if we already knew where x_1 came from, why do we bother to perform the hypothesis test?

Paradox of Conditionality Principle

The conditionality principle has two limitations in reality: (1) The assumption of equal probability when choosing an experiment among m experiments is not realistic. We will almost never choose that way. Instead, we always choose the best type of experiment available for the statistical problem we want to resolve. (2) We often don't know which experiment is performed since, for example, we don't know the value of the parameter. In such cases, the intention becomes a part of the experiment and should be considered in making the inference about the parameter θ. Let me elaborate on this point. When we use the conditionality principle to make a decision; the decision may concern more than one choice (e.g., H_o and H_a), and one of them is performed but others may not be performed. Because the decision-making is related to the unperformed experiments, it is related to potential results from the unperformed experiments. For instance, suppose the single observed value is $x = 2$. If H_1 is true (experiment 1 is performed), the outcomes will mostly be near $\theta_1 = 1.5$, and if H_2 is true (experiment 2 is performed), the outcomes will be mostly near $\theta_2 = 0.5$. As such, we may conclude H_1 is true or prefer H_1 over H_2. On the other hand, if $\theta_1 = 1.5$ and $\theta_2 = 1.9$, the conclusion will be that H_2 is true, based on the single observation $x = 2$.

From the discussions above, it seems the conditionality principle can be generally applied to an estimation problem, but for hypothesis testing the conditionality principle should be used with caution. However, the challenge is that the definition of an experiment is subjective or somewhat arbitrary. For instance, an experiment can be considered as a part of a bigger experiment. Should the conditionality be on the experiment or on the bigger experiment? Such an issue will be encountered again later when we ask, in a sequential experiment with optional stopping, whether the p-value should be calculated conditionally or not.

3.2.2   *Sufficiency Principle*

A statistic T is a function of observations x. The concept of a sufficient Statistic is important in statistical inference. A sufficient statistic contains the whole of the information brought by x about the value of the unknown parameter. Technically, when $x \sim f(x|\theta)$, a statistic T is said to be *sufficient* if the distribution of x conditionally on $T(x)$ is no longer dependent on θ. In other words, if we keep the sufficient statistic $T(x)$ unchanged, the distribution $f(x|\theta)$ will not change as the parameter θ changes.

It is important to know that a sufficient statistic is not for a single distribution with a fixed θ. A statistic is sufficient for a family of probability distributions (e.g., with different values of θ) if the sample from which it is calculated gives no more information than does the statistic as to which of those probability distributions is that of the population from which the sample was taken. Furthermore, when the model allows for a minimal sufficient statistic, that is, for a sufficient statistic which is a function of all the other sufficient statistics, we only have to consider the procedures depending on this statistic. In this way, sufficient statistics can be used for data reduction.

Operationally, the *Fisher–Neyman factorization* is a commonly used tool for finding a sufficient statistic, which can be stated as: A statistic T is sufficient for θ if the density f can be factored into a product, $h \cdot g$, such that h depends on x only, while g depends on x (but only through $T(x)$) and on θ. In other words, if the probability density function is $f(x|\theta)$, then T is sufficient for θ if and only if functions g and h can be found such that

$$f(x|\theta) = h(x) g(T(x), \theta).$$

Example: Binomial Distribution

For a fixed n, denote by $x_1, x_2, \ldots, x_n$ the outcomes of n independent Bernoulli trials with parameter θ ((e.g., flip a coin n times with probability θ of showing heads); then the joint distribution of the random sample is

$$f(x|\theta) = \theta^r (1 - \theta)^{n-r} = h(x) \cdot g(r, \theta),$$

where $h(x) \equiv 1$, $g(r, \theta) = \theta^r (1 - \theta)^{n-r}$, and the number of successes $r = \sum_{i=1}^{n} x_i$. By the Fisher–Neyman factorization theorem, r is clearly a (minimal) sufficient statistic.

The Sufficiency Priniciple states that two observations x and y that are used to construct sufficient statistics $T(x)$ and $T(y)$ such that $T(x) = T(y)$, must lead to the same inference on the model parameter θ.

Here, the term inference refers to hypothesis tests or estimation. There is some confusion about the source of the two observations x and y, that is, do they have to be from the same experimental procedure or can they be from two different procedures (e.g., the experiment with or without optional stopping)? In the frequentist's view, x and y are from the same experimental procedure, but in the likelihoodist's and in the Bayesian view, x and y can be from different experimental procedures. We will elaborate upon this in the Paradox of Birnbaum's Experiment.

Despite the intuitiveness of the Sufficiency Principle, it faces multiple unavoidable challenges as illustrated in the following paradoxes.

Paradox of Sufficiency

In 1934, Fisher discovered the following paradox of sufficiency (Székely, 1986, p.102). Assume $X = (X_1, X_2)$ has a two-dimensional normal distribution whose coordinates are independent with variances 1 and unknown means μ_1 and μ_2. The arithmetical mean $\bar{X} = (\bar{X}_1, \bar{X}_2)$ of the two dimensional sample is a sufficient statistic for μ_1 and μ_2. Suppose the distance from the origin to the point (μ_1, μ_2), $\sqrt{\mu_1^2 + \mu_2^2}$, is known, say, to be 1.5. Then $(\mu_1, \mu_2) = 3(\cos\theta, \sin\theta)$, where the angle θ is the only unknown parameter, but which can be estimated by

$$\hat{\theta} = \tan^{-1}\frac{\bar{X}_2}{\bar{X}_1}.$$

This is an unbiased estimator: $E\left(\hat{\theta}\right) = \theta$ with variance $E(\hat{\theta}-\theta|r = 1.5)^2 = 0.26$. The distribution of $r = \sqrt{\bar{X}_1^2 + \bar{X}_2^2}$ is independent of θ because of its rotational symmetry. Therefore, by the sufficiency principle, no additional information concerning θ can be obtained by taking r into consideration. On the other hand, the efficiency of the estimator $E(\hat{\theta} - \theta)^2$ is very much dependent on r, e.g., $E\left((\hat{\theta} - \theta)^2|r = 3\right) = 0.12$, and $E\left((\hat{\theta} - \theta)^2|r = 4.5\right) = 0.08$.

Fisher's paradox points out that the whole of the information brought by x can be interpreted in different ways. In calculating the efficiency of estimations, the ancillary statistics (like r) may have an important role. This example totally contradicts our general view of sufficiency.

Paradox of Precision with More Data

It is intuitive that more data will enable us to make a better decision or calculate a more precise estimate. The following paradox (Székely, 1986, p. 127), however, shows us this is not necessarily true. Suppose there are four data sources involved with values 0, 1, 2, 2.5. The goal is to select the data whose mean is greatest.

(1) Take samples X_1 and X_2, independently drawn from values 0 or 2 with equal probability 0.5, and X_3 equal to 1 or 2.5 with equal probability. The probability of the correct selection is then

$$P(X_1 < X_3 \text{ and } X_2 < X_3)$$
$$= P(X_3 = 2.5) + P(X_3 = 1)P(X_1 = 0)P(X_2 = 0) = \frac{5}{8}.$$

(2) Now letting the sample size of X_3 increase to 2, the probability of the correct selection becomes:

$$P\left(X_1 < \bar{X}_3 \text{ and } X_2 < \bar{X}_3\right) = 7/16,$$

where $\bar{X}_3$ is the arithmetical mean of the two elements of the sample. Thus, by increasing the sample size, the probability of the correct selection decreases from $5/8$ to $7/16$, or by $3/16$.

Paradox of Information

"Since I started to deal with information theory I have often meditated upon the conciseness of poems; how can a single line of verse contain far more 'information' than a highly concise telegram of the same length. The surprising richness of meaning of literary works seems to be in contradiction with the laws of information theory. The key to this paradox is, I think, the notion of 'resonance.' The writer does not merely give us information, but also plays on the strings of the language with such virtuosity that our mind, and even the subconscious, self-resonate. A poet can recall chains of ideas, emotions and memories with a well-turned word. In this sense, writing is magic." (Székely, 1986).

Thus, information is not a fixed amount, and is dependent on the interaction between the sender and receiver. The quantity of information is measured by the intensity of interaction between the information source and the receiver. It is meaningless to measure the amount of information purely from the source.

3.2.3   *Likelihood Principle*

Likelihood or likelihood function is an important concept in both frequentist and Bayesian paradigms. Let $f(x|\theta)$ be a conditional distribution for X given the unknown parameter θ. For the observed data, $X = x$, the function

$l(\theta|x) = f(x|\theta)$, considered as a function of θ, is called the likelihood function. Here the parameter θ can be viewed as either a nonrandomness variable (frequentism or likelihoodism) or random variable (Bayesianism). Of course, nonrandom is a special case of random with random variability equal to zero.

Similarly, for a discrete probability distribution, the likelihood function is defined as

$$l(\theta|x) = P(X = x|\theta).$$

The likelihood tells us how strongly the data support different values of θ. Therefore, it can be viewed as a function of θ, which measures the relative plausibility of different values of θ. The value of θ corresponding to the maximum value of $l(\theta|x)$ is called the maximum likelihood estimate of θ. The likelihood is not necessarily a probability, but can be a probability density function whose value can be larger than 1.

Note that $l(\theta|x)$ does not have to be a simple closed form; it can be a combination of many different distributional families (e.g., normal, exponential, discrete, continuous). The name likelihood implies that, given x, the value of θ is more likely to be the true parameter than θ_0 is if $l(\theta|x) > l(\theta_0|x)$.

The Likelihood Principle asserts that the information contained by an observation x about quantity θ is entirely contained in the likelihood function $l(\theta|x)$. Moreover, if x_1 and x_2 are two observations, depending on the same parameter θ, such that there exists a constant c satisfying

$$l_1(\theta|x_1) = cl_2(\theta|x_2), \text{ for every } \theta,$$

then they contain the same information about θ and must lead to identical inferences.

When the Likelihood Principle is applied to hypothesis testing, it is only valid when (1) the inference is about the same parameter θ, and (2) the statistic is sufficient under both H_o and H_a.

There are two types of likelihood principles, the weak and strong versions.

Weak Likelihood Principle: According to the Weak Likelihood Principle, the evidential meaning of the statistical evidence (x, M) in an experiment E should depend on the observed data x (but not the model M) only through the observed likelihood function $l(\theta|x)$.

The Weak Likelihood Principle does not imply that evidential meanings in different experiments must be equivalent if the observed data in those experiments have proportional likelihood functions.

Strong Likelihood Principle: According to the Strong Likelihood Principle, the evidential meaning of the statistical evidence (x, M) in an experiment E should depend on the observed data x and the model M only through the observed likelihood function $l(\theta|x)$.

The Strong Likelihood Principle does imply that evidential meanings in different experiments must be equivalent if the observed data in those experiments have proportional likelihood functions.

I should point out that the term "same" in the phrase "the same parameter θ" is not clearly defined, which causes great confusion. The same parameter θ implies that x_1 and x_2 are samples from the same distribution $f(x; \theta)$. However, can x_1 and x_2 obtained from $f(x; \theta)$ be from different sampling methods? If the answer is "yes," then the situation will be equivalent to x_1 and x_2 being obtained from different distributions using the same sampling method; this implies that the Likelihood Principle can be applied to two different parameters, θ_1 and θ_2, as long $l_1(\theta_1|x_1) = cl_2(\theta_2|x_2)$. If the answer is "no," then the Likelihood Principle is a trivial principle since in such cases the two likelihoods are usually identical. Let's use the following paradox to facilitate this discussion.

Paradox of Binomial or Negative Binomial Distribution

Suppose a casino owner claims that his slot game is a fair game, and the reason he can allow this is that as long as he has more money than most gamblers in his casino he still profits. However, we are suspicious of his claim and decided to investigate the losing probability p (in the frequentist's sense of repeated experiments) for a single gambler. We are sure the owner would not make $p < 0.5$. Here is the hypothesis test:

$$H_o : p = 0.5 \text{ vs } H_a : p > 0.5.$$

The experiment was finished with 3 wins out of 12 games. However, from a frequentist's perspective this information is not sufficient for rejecting or accepting the null hypothesis, as discussed below.

Scenario 1: If the total number of games, $N = 12$, is predetermined, the number of losses, X, follows the binomial distribution $B(N; p)$, and the frequentist p-value of the test is given by

$$P(X \geq 9|H_o) = \sum_{x=9}^{12} \binom{12}{x} 0.5^x 0.5^{12-x} = 0.073.$$

The null hypothesis cannot be rejected at a one–sided level, $\alpha = 0.05$. The

likelihood in this case is given by

$$l_1\left(x|p\right) = \binom{12}{9}p^9\left(1-p\right)^3 = 220p^9\left(1-p\right)^3.$$

Scenario 2: If the number of wins, $n = 3$, is predetermined and the gambling stops as soon as 3 wins are observed, then the number of losses, X, follows the negative binomial distribution, $NB(n; p)$, and the frequentist p-value of the test is given by

$$P\left(X \geq 9|H_o\right) = \sum_{x=9}^{\infty}\binom{3+x-1}{2}0.5^x 0.5^3 = 0.0327.$$

Therefore, the null hypothesis is rejected at a one-sided level, $\alpha = 0.05$. The (partial) likelihood in this case is given by

$$l_2\left(x|p\right) = \binom{3+9-1}{2}p^9\left(1-p\right)^3 = 55p^9\left(1-p\right)^3.$$

These two different conclusions are not surprising to any frequentist statistician, because it is a routine hypothesis based on the control of the type I error rate, α.

However, since $l_1\left(x|p\right) = \frac{220}{55}l_2\left(x|p\right)$, we can, according to the strong likelihood principle, conclude that all information is included in $l\left(p\right) = p^9\left(1-p\right)^3$, and the two scenarios should not lead to different conclusions (rejection or not).

HYPOTHESIS TEST & LIKELIHOOD PRINCIPLE

Interestingly, the two scenarios seen in the foregoing example can be restated as taking independent samples from two distributions, the binomial and negative binomial. Thus two different parameters are involved (one could be a desirable response to a drug and the other could be an undesirable response), but the likelihood principle asserts that the two parameters should have the same value (inference). Besides the binomial and negative binomial distributions, we can make many other pairs, for example, with the assistance of a biased coin. For instance, if the result is heads from coin-flipping, we ignore the current result from the NB distribution, and redraw the sample from the NB. Therefore, it is not rational at all to assert that any two parameters from two arbitrary distributions are the same as long as the likelihoods are the same under independent sampling. The likelihoodism paradigm is an axiom system that is inconsistent.

Paradox of Likelihood Principle

The maximum likelihood estimate (MLE) can be biased, for example, the MLE standard deviation σ from a normal distribution $N\left(\mu, \sigma^2\right)$, i.e., $E\left(\hat{\sigma}^2\right) = \frac{n-1}{n}\sigma^2$. The bias can be adjusted by multiplying a factor of $n/(n-1)$. In the example with the negative binomial distribution, the maximum (partial) likelihood estimate of p is $\tilde{p} = \frac{r}{r+k}$, but this is a biased estimate. Its inverse, $(r+k)/r$, is an unbiased estimate of $1/p$, however.

These examples imply that prior knowledge of the bias of a certain estimator can add "information about θ" to the likelihood (i.e., bias correction), so the likelihood does not contain all the information about θ. However, this prior knowledge is independent of the observation x, but related to the distribution of x. If the likelihood does not contain all the information about θ, then there is no sufficient reason to believe the strong likelihood principle.

3.2.4 Law of Likelihood

One criticism of the Neyman—Pearson hypothesis testing is that rejection of the null hypothesis is almost irrelevant to H_a, which is equivalent to always giving the answer "zero" whether you are asked "Is 1 closer to 0 or 10?" or "Is 1 closer to 0 or 1.0001?" This is really difficult to accept. The likelihood ratio test was proposed to overcome the deficiency.

The Law of Likelihood (Hacking 1965) asserts: If one hypothesis, H_o, implies that a random variable X takes the value x with probability $f\left(x|\theta_o\right)$, while another hypothesis, H_a, implies that the probability is $f\left(x|\theta_a\right)$, then the observation $X = x$ is evidence supporting H_o over H_a if $f\left(x|\theta_o\right) > f\left(x|\theta_a\right)$.

The hypothesis testing method that arises directly come from the law of likelihood is the so-called likelihood ratio test. For point-to-point hypothesis testing (3.2), the likelihood ratio is defined as

$$LR = \frac{l(x|\theta_o)}{l(x|\theta_a)}. \tag{3.3}$$

For domain testing (3.1), we can define the maximum likelihood ratio test

$$PM = \frac{\text{upper bound of } l(x|\theta) \text{ when } H_o \text{ is true}}{\text{upper bound of } l(x|\theta) \text{ when } H_a \text{ is true}}. \tag{3.4}$$

When the ratio LR or PM is below a predetermined constant threshold c, we reject the null hypothesis.

Example: Likelihood Ratio Test

Suppose for the hypothesis testing in the Paradox of Binomial or Negative Binomial Distribution, we use the likelihood ratio test based on (3.4). Whether it is scenario 1 (binomial distribution) or scenario 2 (negative binomial distribution),

$$PM = \frac{0.5^{12}}{\max\left(p^3(1-p)^9\right)} = \frac{0.5^{12}}{(1/4)^3(3/4)^9} = 0.20810.$$

If the PM is less than the predetermined threshold c, we conclude that the data support $H_a : p > 0.5$. If it is point-to-point hypothesis testing, for example, $H_o : p = 0.5$ vs: $H_a : p = 0.6$, we can simply use (3.3), so that

$$LR = \frac{0.5^{12}}{0.4^3 0.6^9} = 0.37853,$$

which would also be compared against the constant threshold c to reject or accept H_o.

In the Paradox of Hypothesis Testing, we see the problem when hypothesis testing only concerns the distribution of the null hypothesis. The notion of a likelihood ratio test is that the statistical evidence should be measured based on relative evidence supporting H_o and H_a. Naturally, this method does not guarantee control of the false positive error rate.

When a likelihood ratio test is applied to sequential testing, we call it a sequential likelihood ratio test, which we will discuss later.

Paradox of Law of Likelihood

An estimator $\hat{\theta}(x)$ is a rule for calculating an estimate of a given quantity based on observed data; thus, the rule and its result (the estimate) are distinguished. An estimator is said to be unbiased if the expectation of $\hat{\theta}(x)$

converges to the population parameter θ as the same experiment is repeated, whereas consistency refers to the property of an estimator $\hat{\theta}(x)$ that converges (with probability 1) to the population parameter θ when the sample size increases indefinitely.

A maximum likelihood estimator (MLE) is neither always unbiased nor always consistent. For a finite sample from a normal distribution, the MLE of the standard deviation σ is biased (e.g., for the sample size $n = 2$, $E\left(\hat{\sigma}^2_{MLE}\right) = \frac{1}{2}\sigma^2$), which means that the peaks of the likelihood function indicate (on average) something other than the distribution associated with the drawn sample. As such, how can we say the likelihood is evidence supporting the distribution? Therefore, the likelihood can not be interpreted as the relative plausibility of different θs. If likelihood does not reflect the relative plausibility of different θs, the law of likelihood is losing ground. Furthermore, if K experiments are conducted with sample size n_i $(i = 1, 2, ...K)$ for the ith experiment, the average of the MLEs of the same parameter θ is not necessarily the same as the MLE of θ when all the samples are combined. That is,

$$\frac{1}{K}\sum_{i=1}^{K} MLE\left(n_i\right) \neq MLE\left(\sum_i n_i\right).$$

This seems to further conflict with the statement that the likelihood function contains all the information about a parameter.

Moreover, information about a parameter can have different meanings, just as information about an object can include its weight, size, color, etc. In fact any nontrivial statement about a parameter is information, including an estimate of θ based on x.

3.2.5 Bayes's Law

Bayesian inference is a method of statistical inference in which evidence is used to estimate parameters and predictions in a probability model. In Bayesian inference, all uncertainty is summarized by a posterior distribution, which is a probability distribution for all uncertain quantities, given the data and the model.

The term Bayesian refers to Thomas Bayes, but it was Laplace who first introduced a general version of the Bayesian theorem. Early Bayesian inference is sometimes called inverse probability because it infers backwards from observations to parameters, or from effects to causes. Since the 1980s there has been a dramatic growth in research into and applications of Bayesian methods, mostly because of the discovery of efficient Markov Chain Monte Carlo algorithms.

Human beings ordinarily acquire knowledge through a sequence of learning events and experiences. We hold perceptions and understandings of certain things based on our prior experiences or prior knowledge. When new facts are observed that relate to that prior knowledge we essentially update our perception, or adapt our understanding or develop our knowledge, accordingly. Furthermore, we know that the observed new fact(s) might be more or less likely to reflect the truth. For example, a number of observations that are reasonably consistent might be needed in order to convince us to update our knowledge of some things (e.g., that a particular bus stop is unusually popular along a given road), while a unique, unexpected, large-scale event can be persuasive on other occasions (e.g., commercial aircraft are effective weapons for destroying skyscrapers). But no matter whether the newly observed facts are multiple or solitary, it is this progressive, incremental learning mechanism that is the central idea of the Bayesian approach. Bayes' rule (or theorem) therefore enunciates important and fundamental relationships among prior knowledge, new evidence, and updated knowledge (posterior probability). It simply reflects the ordinary human learning mechanism, and is part of everyone's personal and professional life (Chang and Boral, 2008).

The main difference between the Bayesian approach and the previous two approaches (frequentist and likelihoodist) is that Bayesianism emphasizes the importance of information synergy from different sources: the model $f(x|\theta)$, a prior distribution of the parameters, $\pi(\theta)$, and current data, x. Bayes's Theorem places causes (observations) and effects (parameters) on the same conceptual level, since both have probability distributions. However, the data x is usually observable, while the parameter θ is usually latent.

From Section 2.3, Bayesian Reasoning, Bayes' rule for two hypotheses can be written as

$$P(H_i|D) = \frac{P(D|H_i)P(H_i)}{P(D|H_1)P(H_1) + P(D|H_2)P(H_2)}, \quad i = 1, 2.$$

Defining the Bayes Factor as

$$BF = \frac{P(D|H_1)}{P(D|H_2)} = \frac{P(H_1|D)/P(H_1)}{P(H_2|D)/P(H_2)},$$

we thus obtain from Bayes' rule that

$$\frac{P(H_1|D)}{P(H_2|D)} = BF \cdot \frac{P(H_1)}{P(H_2)}.$$

In layman's terms this says:

Posterior odds ratio = Bayes factor × Prior odds ratio.

In general, the three commonly used hypothesis testing methods in the Bayesian paradigm are:

(1) Posterior Probability: Reject H_1 if $P(H_1|D) \le \alpha_B$, where α_B is a predetermined constant.
(2) Bayes' Factor: Reject H_1 if $BF \le k_0$, where k_0 is a small value, e.g., 0.1. A small value of BF implies strong evidence in favor of H_2 or against H_1.
(3) Posterior Odds Ratio: Reject H_1 if $P(H_1|D)/P(H_2|D) \le r_B$, where r_B is a predetermined constant.

Example: A Diagnostic Test

Bayesian inference based on posterior probability can be illustrated with the following example. A diagnostic test is often used for patient screening before giving any treatment. Denoted by D^+ (D^-) the event with (without) disease; denoted by T^+ (T^-) the event of a (negative) test. Suppose that the sensitivity of a diagnostic test, $P(T^+|D^+)$, is 98%, and the specificity of the test, $P(T^-|D^-)$, is 95%. Thus, $P(T^+|D^-) = 0.05$. We further assume that the disease prevalence (the proportion of the entire population having the disease), $P(D^+)$, is 0.001. Thus, $P(D^-) = 0.999$.

We are now interested in calculating the false positive rate, $P(D^-|T^+)$. From Bayes' rule we reason:

$$\begin{aligned} P(D^+|T^+) &= \frac{P(T^+|D^+)P(D^+)}{P(T^+|D^+)P(D^+) + P(T^+|D^-)P(D^-)} \\ &= \frac{0.98 \times 0.001}{0.98 \times 0.001 + 0.05 \times 0.999} \approx 0.019. \end{aligned}$$

Therefore, the false positive rate is $P(D^-|T^+) = 1 - P(D^+|T^+) = 1 - 0.019 = 0.981$.

Despite the apparent high accuracy of the test, the prevalence of the disease is so low (0.1%) that the vast majority (98.1%) of patients who test positive do not have the disease. Nevertheless, this posterior proportion

$P(D^+|T^+)$ is 19 times the proportion $P(D^+)$, before we knew the outcome of the test!

In the case of a continuous variable x, the posterior distribution is

$$p(\theta|x) = cf(x|\theta)\pi(\theta),$$

where c is a normalization constant and $\pi(\theta)$ is a prior distribution. The Bayes Factor is defined as

$$\text{BF} = \frac{\text{average of } f(x|\theta) \text{ over all values of } \theta \text{ in } H_o}{\text{average of } f(x|\theta) \text{ over all values of } \theta \text{ in } H_a}$$

In other words, the Bayes Factor is defined as the ratio of marginal likelihood of the null hypothesis (the weighted average of $f(x|\theta)$ by the prior $\pi(\theta)$ over all values of θ in H_o) and marginal likelihood of the alternative hypothesis (the weighted average of $f(x|\theta)$ by the prior $\pi(\theta)$ over all values of θ in H_a).

In the Bayesian paradigm, parameter is considered random and the Bayes factor is the marginal likelihood as integrated over all parameters in each model or hypothesis (with respect to the respective priors), whereas in the frequentist and likelihoodist paradigms, θ is not random and for a domain-hypothesis test, a likelihoodist will use the maximum likelihood ratio test (3.4). Therefore, the Bayes factor and likelihood ratio are generally not the same. Nevertheless, when we have a point hypothesis test and probability mass function $P(H_1) = P(H_2)$, the posterior odds ratio, the Bayes factor, and the likelihood ratio are numerically equivalent.

Bayesian inference based on the BF, and in comparison with frequentist hypothesis testing, and likelihood ratio testing, can be illustrated by the following paradox.

Lindley Paradox

Lindley's paradox is about the conflict between conclusions derived from the Bayesian and frequentist approaches. First discussed by Harold Jeffreys (1939), it became known as Lindley's Paradox after Dennis Lindley called the disagreement a paradox (Lindley, 1957). The paradox can occur when the prior distribution is the sum of a sharp peak at the null hypothesis H_o with probability θ and a broad distribution with the remaining probability $1 - \theta$.

Let X be a Bernoulli random variable producing either a success or a failure. Suppose we want to compare a model M_1 where the probability of success is $\theta = 1/2$, and another model M_2 where θ has a prior distribution $\pi(\theta)$ uniformly on [0,1]. We take a sample of 200 from X, and find 115 successes and 85 failures. The likelihood can be calculated according to the binomial distribution:

$$f(x|\theta) = \binom{200}{115}\theta^{115}(1-\theta)^{85}.$$

Therefore, if the sample from model M_1, we have

$$P(X = 115|M_1) = \binom{200}{115}\left(\frac{1}{2}\right)^{200} \approx 0.0059559,$$

and if the sample is from model M_2, we integrate have $f(x|\theta)$ over $[0, 1]$, leading

$$P(X = 115|M_2) \approx 0.004975.$$

From this example we can see the Occam's Razor effect (later in this chapter and in Chapter 4) brought by the Bayes factor: M_1 is a more complex model than M_2 because it has a free parameter which allows it to model the data more closely. The ability of Bayes factors to take this into account is a reason why Bayesian inference has been put forward as a theoretical justification for and generalization of Occam's Razor that penalizes a more complex model and reduces Type I errors (Jeffery and Berger, 1991).

The ratio of the two probabilities is the Bayes factor, which is 1.1972, slightly favoring model M_1. However, if we use a likelihood ratio test, we can determine the maximum likelihood estimate for θ within $[0, 1]$, namely, $f(x|\theta = 115/200) = 0.05699$, and using this gives a likelihood ratio of 0.10451, pointing strongly towards model M_2. More surprisingly, a frequentist hypothesis test of M_1 (H_o) would have produced a more dramatic result, saying that H_o could be rejected at the 5% significance level, since the probability of getting 115 or more successes from a sample of 200 when $\theta = 1/2$ is 0.020, and the p-value for the two-tailed test would be $0.04 < \alpha = 0.05$.

The three approaches, the Bayesian, the likelihoodist, and the frequentist, are in conflict, and this is the paradox.

On the other hand many frequentist statisticians believe the probability of the parameter θ (e.g., treatment effect) being exactly equal to any particular value is zero in most practical situations, therefore, the two-sided test is not meaningful. In other words, the prior belief that $\theta \neq 0.5$ is so strong that no hypothesis test is necessary. Is it ironic that frequentist statisticians have a stronger prior than Bayesian statisticians?

How often does Lindley's Paradox occur in practice? Green and Elgersma (2010) studied the question in ecology to assess the overall severity of the paradox. They randomly sampled p-values and sample sizes reported in the journal "Ecology" in 2009 to estimate the probabilities of acceptance of the null and alternative hypotheses, assuming both were equally likely before observing the data. They found that the average of all significant p-values reported was 0.0083 ± 0.015 (mean $\pm$ standard deviation) and for approximately only 10% of significant (i.e., $p < 0.05$) p-values reported in 2009, the null hypothesis condition was more probable than the alternative. As expected, the difference between the p-value and the probability of a true null hypothesis increased as sample size increased.

Paradox of Posterior Distribution

Suppose we have observation vectors $x_1, x_2, ..., x_n$ from the same model $f(x)$ in n sequential experiments. At the end of each experiment, we calculate the posterior, and such a posterior will serve as the prior for the next experiment. After the nth experiment, we obtain the posterior as $c \sum_{i=1}^{n} f(x_i|\theta) \pi(\theta)$, which represents the updated knowledge. On the other hand, if we deal with the n experiments simultaneously as a single experiment, the posterior distribution will be $cf(x_1, x_2, ..., x_n|\theta) \pi(\theta)$. To be consistent, these two posterior distributions should be the same, but they are usually not unless the $x_1, x_2, ..., x_n$ are independent. More paradoxically, experimental designs (including the type and size of the experiments, and the number of experiments) are often dependent on results of previous experiments (prior knowledge), as in a sequential adaptive design. In other words, the random vectors $x_1, x_2, ..., x_n$ are not necessarily independent or exchangeable. Moreover, it is difficult to determine which previous experiments ought to be considered. This will not only lead to the subjectivity of prior selection but also a logical inconsistency in the Bayesian approach.

3.3 Decision Theory Approach

Note that frequentist hypothesis testing is based on the notion of the type I error rate. The error control method is a worst-case scenario approach, that is, assuming H_o, what is the chance of making a false positive claim? The assumption of H_o may have nothing to do with reality, or the true value of θ. In other words, if we know H_o virtually never happens (see Section 3.4.1, Paradox of the Mixed Experiment), the frequentists have no concern at all and proceed as usual, saying something like "If H_o is true, ..., the probability of observing more extreme data exceeds α, so we cannot reject H_o". This is very irrational from a decisionist point of view. In Decision Theory, we consider a set of possible scenarios, the probability of each, and the impact (loss) if a scenario actually materializes. The probability of a potential event coming to pass can be the posterior probability, calculated from prior and current data. However, the frequentist approach may be considered rational from the game theory perspective, because if the decision-maker does not enforce false positive rate control, people can intentionally change behavior so as to inflate the false positive rate for their own benefit. Such a behavioral change can be viewed as an impact of the decision. Further analysis of such problems will be made in the context of multiplicity in the next chapter.

Statistical analyses and predictions are motivated by objectives. When we prefer one model to an alternative, we evaluate consequences. The potential impact is characterized by a loss function in decision theory. The challenge is that different people have different perspectives on the loss, and hence use different loss functions. Loss functions are often vague and often not explicitly defined, especially when they pertain to decisions we make in our daily lives. Decision theory makes this loss explicit and deals with it with mathematical rigor.

In Decision Theory, statistical models involve three spaces: the sample space X, the parameter space Θ, and the action space A. Actions are guided by a decision rule $\delta(\mathbf{x})$. An action $a \in A$ always has an associated consequence characterized by a loss function $L(\theta, a)$. In hypothesis testing, the action space is $A = \{\text{accept, reject}\}$.

Because it is usually impossible to uniformly minimize the loss $L(\theta, a)$, in the frequentist paradigm a decision rule δ is determined so that the average loss is minimized. An action that minimizes the posterior expected loss is called a Bayes action.

Defining the loss or utility function is controversial and challenging, as illustrated by the following paradox.

3.3.1 *Paradox of Saving Lives*

Some 25 years ago there were heated nationwide debates in China on an ethical problem involving the saving of human life. The issue was idealized as the following catastrophic situation: Your mother, wife and son have fallen into a river and are drowning. None of them can swim. Whom do you save first, knowing that you may not save them all?

First, you love them all equally, so the choice is difficult as you want to save each one. You might think, your mother is weakest and should be saved first. After another second, though, you may want to save your wife first, since without her your child will never be happy. You pause and think again: Of the three, the life expectancy is longest for your child, thus saving the child first is the most meaningful thing to do. But that seems not the best since your wife is a contributor to the workforce and her death would directly impact society the most. Your child is taking the benefits of that society now and is expected to pay back in the future. The mother is living on retirement and social welfare. But would the impact on our lives and society be bigger and outweigh all others in the long term if we adversely discriminate against the elderly? A life is a life regardless of age. Your thinking goes on and on. You struggle and feel paralysis while all your three loves are drwning; it ends with a consequence that may be fair but no one wants it.

You may think I am making a joke about utility or loss measures. Nevertheless, I indeed did research entitled "Impact of diseases and injuries on population," sponsored by The Center for Disease Control (CDC) 15 years ago. There were at least 13 health indices (utility or loss functions) used in practice, including crude death rate, life expectancy, quality of life scores (QOL), QOL-adjusted life-expectancy, and work days lost due to disease and injuries.

3.3.2 *Paradoxes of Group Decisions*

Groups frequently make collective judgments on various propositions. Examples are legislatures, committees, courts, juries, expert panels and entire populations, all deciding what propositions to accept as true, thus forming collective beliefs, and what propositions to make true through their actions, thus forming collective desires. When a group forms collective beliefs, some group members or subgroups may have expert knowledge on certain propositions, and may therefore be granted the right to render judgments upon those propositions (an expert right; Dietrich and List, 2008).

The Expert Panel's Decision

In our society we often delegate our rights to representatives or experts, and let them make the decision for us. This example shows that even if they are more intelligent or knowledgeable than we are, it could still lead to more chances of making mistakes. In the Bayesian decision approach, we often formulate the prior using the opinion of a group of experts. Cautions need to be taken in light of this paradox. Let's cite the example of fallacy concerning the expert right, given by Dietrich and List (2008, without alteration).

An expert committee has to make judgments on the following propositions:

a: Carbon dioxide emissions are above some critical threshold.

b: There will be global warming.

$a \to b$: If carbon dioxide emissions are above the threshold, then there will be global warming.

Half of the committee members are experts on *a*, the other half are experts on $a \to b$. So the committee assigns to the first half the right to determine the collective judgment on *a*, and to the second a similar right on $a \to b$. The committee's constitution further stipulates that unanimous individual judgments must be respected. Now suppose that all the experts on *a* judge *a* to be true, and all the experts on $a \to b$ judge $a \to b$ to be true. In accordance with the expert rights, the committee accepts both *a* and $a \to b$. We may therefore expect it to accept *b* as well. But when a vote is taken on *b*, all committee members reject *b*. How can this happen? The following shows the committee members' judgments on all propositions.

The Liberal Paradox

	a	$a \to b$	*b*
Experts on *a*	True	False	False
Experts on $a \to b$	False	True	False

The experts on *a* accept *a*, but reject $a \to b$ and *b*. The experts on $a \to b$ accept $a \to b$, but reject *a* and *b*. So all committee members are individually consistent. Nonetheless, respecting the rights of the experts on *a* and $a \to b$ is inconsistent with respecting the committee's unanimous judgment on *b*. To achieve consistency, the committee must either restrict expert rights or overrule its unanimous judgment on *b*.

Value of Independent Opinions

Let's discussion the paradox of leadership in the Chapter 1 again, the case of the five-member jury panel with simple majority rule. If the probabilities of making the wrong verdict for members *A*, *B*, *C*, *D*, and *E* are 5%, 10%, 10%, 10%, and 20%, respectively, the probability of bringing the wrong verdict by the panel is about 1%. But if member *E* follows *A*'s verdict, the probability of bringing the wrong verdict will increase to 1.5%. In fact, if the probability of *E* making a wrong verdict is less than 32%, *E* should vote independently; otherwise *E* should follow whatever *A* delivers. In this particular case, 32% is the dividing point where one sees no difference in the outcome whether *E* votes independently or follows *A*.

The Gift of the Magi

"The Gift of the Magi" is a short story, one of several hundred written by O. Henry (William Sydney Porter) between 1903 and 1910. Published in a New York City newspaper in 1905, it is a story about a poor couple who love each other dearly. Della cuts her lovely hair and sells it in order to buy her husband a platinum fob chain for his watch as a Christmas gift. James, meanwhile, sells his watch to buy a comb as his gift to her, since she has beautiful hair. When they come back home each is surprised by what their lover just did. This, of course, is a very moving love story, and the couple are, in the end and in full, emotionally satisfied by their actions. What I want to infer from the story is that two good actions, if taken simultaneously, could lead to an unexpected result. More complex paradoxes, but like this one, are often subjects of Game Theory, and we will discuss an example or two in Chapter 4.

3.4   Controversies in Evidence Measures

3.4.1   *Paradox of Model Fitting and Validation*

Authors frequently write in their prefaces that there will inevitably be errors in the book. If what they write is true, there will be at least one false statement in the book; otherwise, the prefatorial claim is false. Either way, they are committed to a falsehood, and must be guilty of an inconsistency (Clark, 2007, p. 167; Sainsbury, 1995, p. 148; Prior, 1961, p. 85).

Similarly, as we do model checking, when extreme samples do occur the model validation procedure will reject the true model. At the same time model validation can also lead to the inclusion of data from a different model. Thus, by doing such model checking and under a repeated experiment scenario, all the models will be truncated and/or mixed with each other. This further implies we invalidate all models by performing model validation. Therefore, based on repeated experiments, the type I error may not be controlled. On the other hand, if a wrong model can be used without model validation, the type I error may again not be controlled. Either way, we don't have the right model.

3.4.2   *Robbins's Paradox*

On the basis of a single observation x, we know that x itself is a minimum variance, unbiased, maximum likelihood estimator for the parameter of a Poisson distribution. Given that, we naturally think that, for the parameters θ_1, θ_2,...,

θ_n of n independent Poisson distributions, on the basis of the corresponding observations x_1, x_2, ..., x_k, with x_i from the model with θ_i, $\hat{\theta}_i = x_i$ would be the best estimator that minimizes

$$E\left(\sum_{i=1}^{k}\left(\hat{\theta}_i - \theta_i\right)^2\right).$$

However, H. Robbins pointed out that (Székely, 2008, p. 132—133), though the k Poisson distributions are independent and may represent complete events, it is still possible to find better estimators which take into account not only their own observations, but also those of the others. If k is large and $N(X)$ denotes the number of observations which are equal to X, Robbins shows that the estimator

$$\hat{\theta}_i = (x_i + 1)\frac{N\left(x_i + 1\right)}{N\left(x_i\right)}$$

is better than $\hat{\theta}_i = x_i$. This paradox implies that it is possible that observations which have nothing to do with a parameter can influence the goodness of its estimates.

Similarly, Charles Stein's paradox (Stein,1956; Efron and Morris, 1977) is the phenomenon that when three or more unrelated parameters are estimated simultaneously, there exist combined estimators more accurate on average (that is, having a lower expected mean-squared error), than any method that handles the parameters separately.

To illustrate the counter-intuitive nature of Stein's example, consider the following real-world example. Suppose we are to estimate three unrelated parameters, such as the performance score of a randomly selected school in Washington, DC, the number of visitors to China each year from 2000 to 2010, and the average weight of an American. The above example tells us that we can get a better estimate (on average) for the vector of three parameters by simultaneously using three unrelated measurements. However, it does not mean that we get a better estimator for the performance score by measuring some other unrelated statistics such as the number of visitors to China or the average weight of an American. We have not obtained a better estimator for the performance score itself, but we have produced an estimator for the vector of the means of all three random variables, which has a reduced total error. So the cost of a worse estimate in one component might be compensated by a better estimate in another component.

3.4.3 *Simpson's Paradox*

In probability and statistics, Simpson's Paradox, introduced by Colin R. Blyth in 1972, points to apparently contradictory results between aggregate data analysis and analyses from data partitioning. Let's look into a few real-life examples.

Berkeley Sex Bias Case

One of the best known real-life examples of Simpson's paradox occurred when the University of California, Berkeley, was sued for bias against women who had applied for admission to graduate schools there. The admission rates (44% for men and 35% for women) for the fall of 1973 showed that men applying were more likely than women to be admitted, and the difference was statistically significant (Bickel, Hammel, and O'Connell, 1975). However, when examining the individual departments, it was found that most departments had a "small but statistically significant bias in favor of women."

The authors conclude: "Examination of aggregate data on graduate admissions to the University of California, Berkeley, for fall 1973 shows a clear but misleading pattern of bias against female applicants. Examination of the disaggregated data reveals few decision-making units that show statistically significant departures from expected frequencies of female admissions, and about as many units appear to favor women as to favor men. If the data are properly pooled, taking into account the autonomy of departmental decision making, thus correcting for the tendency of women to apply to graduate departments that are more difficult for applicants of either sex to enter, there is a small but statistically significant bias in favor of women. The graduate departments that are easier to enter tend to be those that require more mathematics in the undergraduate preparatory curriculum."

Kidney Stone Treatment

Charig et al. undertook a historical comparison of success rates in removing kidney stones. Open surgery (1972—1980) had a success rate of 78% while percutaneous nephrolithotomy (1980—1985) had a success rate of 83%. However, the success rates when considered for each group (small and large stones) are surprisingly better for open surgery than for percutaneous nephrolithotomy (Charig et al., 1986, Julious and Mullee, 1994).

The practical significance of Simpson's paradox surfaces in decision-making situations, where it poses the following dilemma: Which data analysis should we consult in choosing an action, the aggregated or the partitioned? In the kidney stone example above, it is clear that if one is diagnosed with "small stones" or "large stones" the data for the respective subpopulation should be consulted and open surgery would be preferred. But what if a patient is not diagnosed and the size of the stone is not known? Would it be appropriate to consult the aggregated data and perform the nephrolithotomy? This would stand contrary to common sense; a treatment that is preferred both under one condition and under its negation should also be preferred when the condition is unknown (wikipedia.org).

Kidney Stone Treatment Outcomes

	Open surgery	Percutaneous nephrolithotomy
Small stone	93% (81/87)	87% (234/270)
Large stone	73% (192/263)	69% (55/80)
Total	78% (273/350)	83% (289/350)

As of early 2008, there were 50,629 clinical trials ongoing globally, up by 1.3% over 2007. Clinical trials across multiple regions of the world have become common practice. If very different drug effects were observed in different countries, how should the drug be used in different countries? A second question is: Should the regions surveyed be states, cities, or arbitrary regions?

If the partitioned data is to be preferred a priori, what prevents one from partitioning the data into arbitrary subcategories artificially constructed to yield wrong choices of treatments?

In the sex bias case, if we keep hunting for a positive finding, we will eventually find cases of sex bias against women. However such a finding could be a false positive because of the so-called multiplicity issue (see the next chapter).

3.4.4 Regression to the Mean

Regression to the mean was first identified by Sir Francis Galton in the nineteen century. He was a half-cousin of Charles Darwin, a geographer, meteorologist, and tropical explorer, a founder of differential psychology, the inventor of scientific fingerprint identification, a pioneer of statistical correlation and regression, a convinced hereditarian and eugenicist, and a best-selling author.

Sir Francis Galton discovered the phenomenon that sons of very tall fathers tended to be shorter than their fathers and sons of very short fathers tended to be taller than their fathers. Another similar phenomenon is noticed: A class of students takes two editions of the same test on two successive days; it has frequently been observed that the worst performers on the first day will tend to improve their scores on the second day, and the best performers on the first day will tend to do worse on the second day. This phenomenon is called "regression to the mean," and is explained as follows. Exam scores are a combination of skill and luck. The subset of students scoring above average would be composed of those who were skilled and had not especially bad luck, together with those who were unskilled but were extremely lucky. On a retest of this subset, the unskilled will be unlikely to repeat their lucky break, while the skilled will have a second chance to have some pretty bad luck. Hence, those who did well previously are unlikely to do quite as well in the second test.

The phenomenon of regression to the mean holds for almost all scientific observations. Thus, many phenomena tend to be attributed to the wrong causes when regression to the mean is not taken into account. What follows are some real-life examples.

The calculation and interpretation of "improvement scores" on standardized educational tests in Massachusetts probably provides an example of the regression fallacy. In 1999, schools were given improvement goals. For each school the Department of Education tabulated the difference in the average score achieved by students in 1999 and in 2000. It was quickly noted that most of the worst-performing schools had met their goals, which the Department of Education took as confirmation of the soundness of their policies. However, it was also noted that many of the supposedly best schools in the Commonwealth, such as Brookline High School (with 18 National Merit Scholarship finalists), were declared to have failed. As in many cases involving statistics and public policy, the issue was debated, but "improvement scores" were not announced in subsequent years, and the findings appear to be a case of regression to the mean (wikipedia.org).

The psychologist Daniel Kahneman, winner of the 2002 Nobel Prize in Economics, pointed out that because we tend to reward others when they do well and punish them when they do badly, and because there is regression to the mean, it is part of the human condition that we are statistically punished for rewarding others and rewarded for punishing them (Daniel Kahneman's autobiography is at nobelprize.org).

Recent United Kingdom law enforcement policies have encouraged the visible setting of static or mobile speed cameras at accident blackspots. This policy was justified by a perception that there is a corresponding reduction in serious road traffic accidents after a camera is set up. However, statisticians have pointed out that, although there is a net benefit in lives saved, failure to take into account the effects of regression to the mean results in the beneficial effects being overstated. It is thus claimed that some of the money currently spent on traffic cameras could be more productively directed elsewhere (wikipedia.org).

Paradoxes of Regression to The Mean

Mr. Lee is a very tall man in his town, so his son is likely shorter than he is. Paradoxically, outside his town Lee is considered a very short man (in his country), so his son is more likely taller than he is. Which is true?

A student known for his math skill is seen to be at the top of his class for many years. Yesterday he took a math test. His result was above the average for the class, but the worst among his math scores for the past several years. Considering regression to the mean, does it mean he is likely to do worse or better on tomorrow's test? This is an issue of causal space, as we discussed in Chapter 2 and will be discussed more later in this chapter.

For the case of improvement scores on standardized educational tests in Massachusetts, suppose the average score of a school over time is the same as the average score of all the schools. (i.e., the time-average = the space-average. Such a condition is called an ergodic condition. Brownian motion is ergodic.) Then, interestingly enough, any part of such a "flow" (the average score of some of the schools changes over time, and can be viewed as the flow of a liquid) has a memory that deflates the contribution of past values, as we discussed in Chapter 2 apropos the Assuan Dam.

3.4.5 *Paradox of Confounding*

A confounding variable (also confounding factor, lurking variable, confounder) is an extraneous variable in a statistical model that correlates positively or negatively with both the dependent and independent variables. Whereas bias involves error in the measurement of a variable, confounding involves error in the interpretation of what may be an accurate measurement. A classic example of confounding is to interpret the finding that people who carry matches are more likely to develop lung cancer as evidence of an association between carrying matches and lung cancer. Smoking is the confounding factor in this relationship: Smokers are more likely to carry matches, and they are also more likely to develop lung cancer. However, if "carrying matches" is replaced with "drinking coffee," we may easily conclude that coffee more likely causes cancer. Interestingly enough, epidemiologic studies have evaluated the potential association between coffee consumption and lung cancer risk. The results were inconsistent. Recently Tang et al. (2010) conducted a meta-analysis of eight case-control studies and confirmed that coffee consumption is a risk factor for lung cancer.

A factor is not a confounder if it lies on the causal pathway between the variables of interest. For example, the relationship between diet and coronary heart disease may be explained by measuring serum cholesterol levels. Cholesterol is not a confounder because it may be the causal link between diet and coronary heart disease. Bias creates an association that is not true, while confounding describes an association that is true, but potentially misleading (collemergencymed.au.uk). It is a common view that the confounding issue can be handled by regression. However, such a view is challenged by our next paradox.

Suppose seven subjects in each of two treatment groups (A and B) are measured for clinical and biomarker responses. We first calculate Pearson's correlations between the variables: the treatment, biomarker, and clinical endpoints. The results are presented in the figure below. We can see that the correlations between them are not transitive. In other words, a correlation between the treatment and the biomarker ($R_{XB} = 0.45$) and a correlation between the biomarker and the clinical responses ($R_{BC} = 0.90$) do not ensure a correlation (R_{XC}) between the treatment and the clinical response.

The average response with the clinical endpoint is 4 for each group, which indicates that there is no treatment effect. On the other hand, the average biomarker response is 6 for group B and 4 for group A, which indicates that the drug has effects on the biomarker.

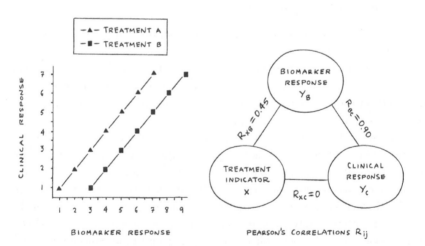

After fitting a linear model with the data we obtain

$$Y_C = Y_B - 2X.$$

This model fits the data well based on the model-fitting p-value and R^2 (the proportion of variability explained by the model). Specifically, R^2 is equal to 1. The p-values for the model and all parameters are equal to 0. The coefficient 2 in the model is the separation between the two lines. Based on the model, we would conclude that both the biomarker and the treatment affect the clinical response. But in fact the treatment has no effect on the clinical response at all. The reason is that Y_B relates to X by $Y_B = 2X$. This implies the $Y_C = 2X - 2X = 0 \cdot X$. We can even create an artificial variable, e.g., $Y_B = 3X$ or $Y_B = -X$, which changes the treatment effect, X, in the model. Indeed, due to the correlations between the explanatory variables and the subjectivity in selecting the variables, the regression result is difficult to interpret. "Statistics can be made to prove anything—even the truth."

This phenomenon, that adding explanatory variables may reverse the relationship between the outcome variable and explanatory variables, is called Lord's paradox by some scholars (Lord, 1967; Tu, Gunnell, and Gilthorpe, 2008).

3.5   Causal Space Theory: Unification of Paradigms

The Principle of Evidential Totality: In any experiment, the information about a parameter is always more than the observations alone. This extra information includes the prior knowledge about the unknown parameter θ, the choice of experiment design, and hypotheses that should be relevant in making an inference about θ.

A sufficient statistic T adapts to both the null and alternative hypotheses, because T is sufficient for the family of distributions under both hypotheses. However, such a consideration may not be enough. For instance, suppose T is sufficient for the normal distribution family $N(\theta, 1)$ and $N(0, 1)$ is the distribution under the null hypothesis. Then T will not differentiate $N(1, 1)$ or $N(100, 1)$ for the alternative hypothesis.

The prior knowledge of θ is only partially reflected in the hypotheses and experimental design, e.g., the hypotheses reflect the possible range of θ, the experimental design (including the intention of the designer), and may or may not reflect the uneven probabilities of selecting different θs. How the difference in hypotheses will affect the inference on θ has been discussed in Section 3.2.1, in the paradox of the conditionality principle.

3.5.1 *Paradox of Mixed Experiment*

Consider a mixed experiment with two steps. In Step 1, randomly select $\theta = 0$ and $\theta = 1$ with probability p and $(1 - p)$, respectively. In Step 2, after θ is selected, a single random sample x is drawn from the distribution $f(x|\theta)$. Suppose you are given p and x (but θ is hidden or not observable) and asked to make an inference on θ, the source of x.

To make an inference on θ, both p and x should be used. Let me elaborate what I mean by this through some examples.

Suppose we have $p = 1 - 0.1^{1000}$, $x = 1.645$, and $f(x|\theta) = N(\theta, 1)$. Using the frequentist hypothesis approach with the test statistic $T(x) = x$ (it is a sufficient statistic because it's a single sample) and the significance level $\alpha = 0.05$, we reject $\theta = 0$. However, given $p = 1 - 0.1^{1000}$, we are pretty sure x is from $f(x|\theta = 0)$. In other words, under the (infinite) repetition of the mixed experiment, $x = 1.645$ (or $x \geq 1.645$) is more often generated from the source with $\theta = 0$ than with $\theta = 1$.

Therefore, $f(x|\theta)$ is reflected in the two hypotheses, p being the prior, and x being the observation, all contributing to the information about θ. We should not ignore any of them when we do our inference on θ. Furthermore, the contributions of $f(x|\theta)$, p and x to the information about the unknown source parameter θ come in a very complicated way, or at least not simply as an additive function. For instance, without x, we can say "$P(\theta = 0) = p$ and $P(\theta = 1) = 1 - p$." After x is observed, what can we say about θ? We may say "$P(\theta = 0) = p^*$ and $P(\theta = 1) = 1 - p^*$," where p^* is a new value related to x. Hence, the difference between p^* and p characterizes the information brought by x solely. However, in general, such an informational difference (a single value) can be defined in several ways. In information science, information is measured by entropy; in computer science, information is measured by complexity, which is related to randomness.

Let's now discuss the Likelihood Ratio Test. It is obvious that the test, or the law of likelihood, has the same serious problem as frequentist hypothesis testing, because it fails to consider the random selection probability p.

For the Bayesian approach we generally consider the prior, but different measures use the prior differently. The posterior distribution and posterior odds ratio incorporate the prior distribution of p, but the Bayese Factor does not, as illustrated by the Lindley Paradox.

In reality, the random selection probability p is not known. For example, let's denote by $\theta = 0$ the presence of an ineffective chemical compound and by $\theta = 1$ that of an effective compound in drug development. Then p can be the proportion of ineffective compounds. If based on the repetition of the mixed experiment, x has the mixed distribution

$$g(x; p) = pf(x; \theta = 0) + (1 - p) f(x; \theta = 1).$$

This mixed distribution can be used to estimate the proportion of ineffective compounds.

It is important to know that not all five statistical principles are mutually independent. The next paradox, the Birnbaum experiment, will show you why.

3.5.2 Paradox of the Birnbaum Experiment*

The argument from the Birnbaum Experiment purports to derive the strong likelihood principle (SLP) from sufficiency and conditionality principles (Birnbaum, 1965; Berger and Wolpert, 1988; Casella and Berger, 2002). Let's outline Birnbaum's arguments.

If two observations y'^* and y''^* are from two experiments E' and E'', respectively, and have proportional likelihoods, then they have a common sufficient statistic $T(y'^*) = T(y''^*)$, since the likelihood itself is a sufficient statistic. Furthermore, the SLP states that the inference about parameter θ based on y'^* from E' should the same as that based on y''^* from E''.

$$\Upsilon_{E'}(y'^*) = \Upsilon_{E''}(y''^*),$$

where Υ means "inference."

We now consider a mixture experiment wherein a coin (fair or biased) is flipped, with "heads" leading to performing E' and reporting the outcome y', and "tails" leading to E'' and reporting the outcome y''. Each outcome would have two components (E^j, y^j) $(j = '$ or $'')$, and the distribution for the mixture would be sampled over the distinct sample spaces of E' and E''.

In the Birnbaum experiment there are two possible cases in terms of the outcomes:

- Case 1, when E' and E'' have "star paired" outcomes (i.e., their likelihoods are proportional), we define the test statistic using the common sufficient statistic $T(y'^*)$. In other words, we always report $T(y'^*)$, even if the data y''^* are actually observed.
- Case 2, when E' and E'' don't have "star paired" outcomes (i.e., their likelihoods are not proportional), we formulate the test statistic (usually) based on the experiment performed E^j and outcome y^j, where $j = '$ or $''$.

That is, Birnbaum's experiment, E_{BB}, is based on the statistic T_{BB}, given by

$$T_{BB}(E^j, y^j) = \begin{cases} T(y'^*) & \text{if } j = 1 \text{ and } y' = y'^* \text{ or if } j = 2 \text{ and } y'' = y''^* \\ T(E^j, y^j), & \text{otherwise} \end{cases}$$

Because the argument for the SLP is dependent on Case 1 outcomes, we may focus only on them for now. Let's start with the following two premises:

(1) Because there is a common sufficient statistic $T_{BB}(E^j, y^j) = T(y'^*)$ for E' and E'', based on the sufficiency principle, we have the first premise of the argument:

$$\Upsilon_{E_{BB}}(E', y'^*) = \Upsilon_{E_{BB}}(E'', y''^*). \tag{3.5}$$

(2) The argument next points out that the conditionality principle tells us that, once it is known which of E' or E'' produced the outcome, we should compute the inference just as if it were known all along that E^j would be performed. Applying the conditionality principle to Birnbaum's mixture gives the premise:

$$\Upsilon_{E_{BB}}(E^j, y^{j*}) = \Upsilon_{E^j}(y^{j*}). \tag{3.6}$$

Expanding (3.6) into two equations for $j = '$ and $''$ and using (3.5), we immediately have $\Upsilon_{E'}(y'^*) = \Upsilon_{E''}(y''^*)$, that is, the inference from y'^* is identical to the inference from y''^*, which is the SLP.

Paradoxically, Professor of Philosophy Deborah Mayo (Mayo and Spanos, 2010, p. 305—314) claims to have disproved Birnbaum's argument by pointing to an apparently hidden fault in the proof. Mayo states: "The problem is that, even by allowing all the premises to be true, the conclusion could "follow" only if it is assumed that you both should and should not use the conditional formulation. The antecedent of premise (1) is the denial of the antecedent of premise (2)."

However, Mayo's disproof is faulty because her presumption about the antecedent of premise (1) in Birnbaum's argument is odd. A sufficient statistic is sufficient for a FAMILY of distributions with different values of the parameter θ; such a family of distributions often consists of the distributions under H_o and H_a. Remember, given a sufficient statistic T, the distribution of x will vary as θ varies. Therefore, her statement about the sufficient statistic under a mixed distribution (a fixed distribution) is irrelevant.

Frequentists don't believe the likelihood principle because they don't believe conditionality, as we discussed in Section 3.2.1.

3.5.3 *Paradoxes of Causal Space*

When we talk about an experiment, we usually do not define it well; in other words, we do not define the causal space. The casual space is a hidden assumption. For this reason, different paradigms imply different causal spaces for the "same" experiment and thus have different outcome spaces. Different causal spaces lead to controversial results, as illustrated in our next set of paradoxes.

Paradox of The Traffic Ticket

Let me share my personal story with you. When I appeared in court for my very first traffic citation, I argued: "I don't believe this my fault... I don't have any traffic violations in my 13 years of driving." In Bayesian terms, I was saying I have a very strong prior record among all car accidents in which I was involved, none of them was my fault; therefore, this time it is very likely not my fault either. However, the judge replied nicely, "Yes I know, but everyone has a first time". Her prior is quite different from mine. She probably thought that the (unconditional) probability of a driver who had a 13-year driving history with at least one traffic violation is very high. The prior is subjective and depends on how the individual uses his knowledge. In the current case, we can even use traffic records from all people who have similar characteristics (age, gender, driving history, etc.) to construct the prior; but the word "similar" here is very subjective. However, do you think it is fair to infer what I did in terms of what others have done (the prior)? On the other hand, isn't science used in general to make inferences based on prior knowledge or natural laws?

Paradox of Life Expectancy

A Chinese man was taking a trip from China to Japan. When landed in Japan, he announced, excitedly, that he had just increased his life expectancy by 11.3 years, since the life expectancy is 86.1 years in Japan and 74.8 years in China. Obviously his causal space is based on where he is at the moment of reckoning. When he was in China, he considered himself Chinese and calculated the life expectancy based on the causal space of all Chinese; when he was in Japan, he considered himself Japanese and calculated the life expectancy based on the causal space of all Japanese. But (speaking for myself) I don't remember feeling any younger when I took the oath at the Naturalization

Oath Ceremony in the Massachusetts State Hall and became a U.S. citizen!

Paradox of Drug Effects

Suppose there are two random variables X and Y, representing, for example, the treatment effects of a drug for two populations, A and B. Now the problem is to infer the drug's effect (denote by a fixed value $\hat{Z}$) on me. If we put $\hat{Z}$ with X, we formulate a new population G; if we put $\hat{Z}$ with Y we have the new population H. Because A and B are very large populations, including or not including $\hat{Z}$ will not affect overall the conclusions about drug effects on those populations. Therefore, if the drug has a positive effect on A it will also have a positive effect on G; likewise if the drug has a negative effect on B, it will have a negative effect on H. Now the question is, should I draw a conclusion about the drug's effect on me based on population G or H? Remember, I don't have to have the same identified characteristics as A or B since populations G and H are defined by the aggregated properties of A + me and B + me, respectively.

Paradox of Stock Price Trends

When we observe stock prices steadily increasing for a long time, our confidence increases that the trend will continue. On the other hand, the longer the trend continues, the more likely the market will drop soon. Which of the two is true depends upon the causal space. Such a paradox can be applied to a wide variety of scientific laws. However, suppose the second belief is true. Then the statement "the longer the trend continues, the more likely the natural law will break" itself becomes more and more likely to break as time goes by—a form of self-referential paradox.

Paradox of Sampling Independent Inference

If you believe anything happens (e.g., action is taken) for a reason, then samples may never be independent, else there would be no randomness. Just as T. Hilberman put it (Robert, 1994): "From where we stand, the rain seems random. If we could stand somewhere else, we would see the order in it."

3.5.4   *Causal Space Schema of Statistical Paradigms*

All different statistical paradigms can be unified in terms of causal space schema, that is, different paradigms use different causal space schema. The sampling procedure or other aspects relevant to the experiment will guide us as to how to pool "similar" situations or events to form the "causal space" for making inferences. In Chapter 2, in discussing the concept of probability, we illustrated how frequentism determines such similarity explicitly, whereas Bayesianism determines such similarity implicitly, especially for subjective Bayesianism.

It is critical to realize that when we discuss inference or information about a parameter θ, we treat θ as a (random or non-random) variable, i.e., we are working only on our knowledge space of θ, because θ itself is a fixed number. When we view θ as a fixed but unknown value, we in fact treat θ as a variable (random or nonrandom). Such a view is to consider θ in the context of a collection of similar situations. The notion of causal space schema of different paradigms is built on this view of probability and the knowledge space of θ.

Causal Space Schema of Frequentism

The frequentist's causal space depends on the experimental procedure. For instance, the causal space will be different for an experiment with a fixed sample size and a sequential experiment with optional stopping, as we have seen in the Paradox of Binomial or Negative Binomial Distribution.

Causal Space Schema of Likelihoodism

A likelihoodist's causal space does not depend on the experimental procedure. For instance, the causal space will be the same for an experiment with a fixed sample size or with sequential stopping, as we have seen in the Paradox of Binomial or Negative Binomial Distribution. Thus, the population has a mixed distribution of binomial and negative binomial distributions. However, the determination of the causal space is difficult because we don't know the proportion of each procedure. It is even more challenging if the selection of future experimental procedures is dependent on the chosen statistical paradigm (frequentist hypothesis test, likelihood ratio test, or a Bayesian approach).

Causal Space Schema of Bayesianism

Prior is a subset of the causal space. Choice of a prior should depend on the experimental procedure, because different procedures will lead to different false positive rates. For instance, the current statistical criterion for new drug approval is based on a type I error control at the 2.5% level. If it were based on the likelihood principle, there would be more false positive drugs, thus the prior would be a more skeptical prior.

Suppose we want to project the probability of success of the current drug and the probability of effectiveness of the current drug given the data in the current trial. Furthermore, suppose the previous data have grouped into two categories: (1) the test drugs investigated using a fixed sample size, the proportion of success among them is 10%, (2) the test drugs conducted using a sequential trial without adjusting α (hypothetically is acceptable to the regulatory agency in charge of drug approval for marketing), the success rate being 15%. We also know of no reason to believe that the proportion of effective drugs should be different in the two groups. The only explanation for the different rates is that the sequential method inflates the false positive rate. The question is: to estimate the probability of success for the current drug, how should one determine the prior? In the spirit of conditionality, if the current study uses a sequential design, we should use the 15% success rate as the prior; otherwise we should use 10%. However, the likelihood principle and Bayesian statistics don't differentiate the procedures, and thus the prior will be the weighted average of 10% and 15%, regardless of the current trial design. On the other hand, we can reason that if sequential designs were used for all previous trials, the prior success rate would be 15%, and therefore 15% should be used for the prior. In contrast, we can also reason that if fixed sample size designs were used, the success rate would be 10%, so 10% should be the prior. Which is correct?

The similarity controversy also arises when estimating the probability of effectiveness of current drugs.

This discussion also naturally leads to the issue of multiplicity or multiple-testing problem, which is our next topic.

3.6   Multiplicity: The Black Hole of Scientific Discovery

Scientists conduct observational and prospective experiments since they believe there are causal relationships in the world. They propose hypotheses and hope to prove or disprove them through hypothesis testing. Science is nothing but an attempt to make conclusions based on observations, experiments, and analysis. We have seen that controversies and paradoxes surrounding scientific inference. In this chapter, we are going to discuss a very challenging issue in quantifying scientific evidence, i.e., multiplicity. Generally, the word *multiplicity* refers to the issue of aggregate error control when multiple hypothesis tests or analyses are conducted.

Multiplicity exists in every aspect of science and of our daily lives, just like the black hole that attracts everything. Multiplicity is the horizon where all different statistical paradigms merge. At the same time, multiplicity might be a keyhole leading us to the resolution of philosophical differences among these same statistical paradigms.

3.6.1   Description of Multiplicity Issues

Paranormal Coincidences

In a certain national lottery a few years ago, there was a single winner. Ten million \$1 lottery tickets were issued, and there was one and only one winning ticket. Ten million people each bought one lottery ticket. Thus, each of them had a very small chance of winning (1 out of 10 million). Mr. Brown won the lottery, but he also got a lawsuit charging him with conspiracy. The prosecution argued that the chance of Mr. Brown winning the lottery was so low that it could not happen unless there were a conspiracy. However, Mr. Brown defends: "There must be a winner whoever he/she might be, regardless of any conspiracy theory." This paradox is a typical multiplicity problem. The multiplicity emerges when we apply the conspiracy theory (hypothesis test) to each of the ten million lottery buyers. A multiple hypothesis test (there is a conspiracy) will lead to a false conclusion with a larger probability.

Similarly, in our justice system, if in each case there is a very low probability of wrongly convicting someone (even using DNA technology), there are still many people punished for crimes they did not commit.

PARANORMAL COINCIDENCE

Let's introduce the example given by Vidakovic (2008): Suppose a burglary has been committed in a town, and 10,000 men in the town have their fingerprints compared to a sample from the crime. One of these men has a matching fingerprint, and at his trial it is testified that the probability that two fingerprint profiles match by chance is only 1 in 20,000. This does not mean the probability that the suspect is innocent is 1 in 20,000. Since 10,000 men had fingerprints taken, there were 10,000 opportunities to find a match by chance; the probability of at least one fingerprint match can be calculated from the binomial distribution, i.e., $1 - \binom{10000}{0} \left(\frac{1}{20000}\right)^0 \left(1 - \frac{1}{20000}\right)^{10000} \approx 39\%$, which is considerably more than 1 in 20,000.

Here is another real life example. Sally Clark was a British solicitor who became the victim of a famous miscarriage of justice when she was wrongly convicted of the murder of two of her sons in 1999. Clark's first son died suddenly within a few weeks of his birth in 1996. After her second son died in a similar manner, she was arrested and tried for the murder of both sons. A "statistical" claim against her was that the probability of two children from an affluent family suffering sudden infant death syndrome was 1 in 73 million. The Royal Statistical Society later issued a public statement expressing its concern at the "misuse of statistics in the courts" and arguing that there was no statistical basis for such a claim. The convictions were overturned in January 2003, after it emerged that the prosecutor's pathologist had failed to disclose microbiological reports suggesting that one of her sons had died of natural causes. She was released from prison having served more than three years of her sentence.

There are plenty of other examples of multiplicity problems from daily life. Many people prognasticate concerning economic trends, the housing market, or earthquakes, and many of them will guess right. Ironically, if you are "nobody," no one cares about and no one knows your prediction. On the other hand, if you are a Wall Street guru or a well-known scientist, your guess will be characterized as an "intellectual" prediction.

Multiplicity in Scientific Discovery

Multiplicity is a commonly faced problem in drug development. A simple example of multiplicity would be as follows. When we test the effectiveness of a placebo or ineffective compounds in patients thousands of times, we will eventually see that the placebo works for some patients. If we see a sufficient (fixed) number of patients responding to the placebo, we stop the test and claim the placebo works for the underlying disease. Everyone would agree that this can be a problem, i.e., leading to a false positive claim. Thus, frequentists have developed a hypothesis test to control the type I error (false positive claims), and such error control is enforced within an experiment through so-called experiment-wise (or familywise) error control.

Multiplicity can come from varying sources, for example, in clinical trials: (1) multiple-treatment comparisons, (2) multiple tests performed at different times, (3) multiple tests for several clinical endpoints (parameters), (4) multiple tests conducted for multiple populations using the same treatment within a single experiment, and (5) sequential tests in so-called adaptive design trials.

Other examples of multiplicity issues would be: (1) In analysis of gene expression microarrays, we are interested in the mean differential expression, μ_i, of genes $i = 1, ..., 10,000$, using hypothesis tests: $H_o : \mu_i = 0$ versus $H_a : \mu_i \neq 0$. The multiplicity problem is that even if all $\mu_i = 0$, one would find that roughly 500 tests reject at, say, the level $\alpha = 0.05$, and a correction for this effect is needed. (2) Pharmacovigilance involves monitoring drug safety by conducting a sequence of analyses on a cumulative adverse event database. Without multiplicity control all drugs will eventually be flagged as unsafe. (3) In compliance with guidelines promulgated in the document Synchronic Surveillance of the National Defense and Homeland Security, many counties in the U.S. perform daily tests on the "excess" of some medical symptoms, wishing to detect early any outbreak of epidemics or bio-terrorist attacks. Such tests can result in a large number of false discoveries without multiplicity adjustment (Berger 2009).

Multiple-hypothesis testing can inflate the type I error dramatically, without proper adjustments for the p-values or significance level. Suppose we perform two independent hypothesis tests at a significance level of 0.05, so that the probability of rejecting one of the null hypotheses will be larger than 0.05. Such an increase in rejection probability is called error inflation. Given two hypothesis tests, the level of inflation is dependent on the correlation between the two test statistics. The maximal inflated error rate (0.098) occurs when the endpoints are independent. If the hypotheses are perfectly correlated, there is no error inflation. For a correlation as high as 0.75, the error rate is still larger than 0.08. Hence, to control the overall type I error rate, α, an alpha adjustment is required for each test. The error-rate inflation is also related to the number of independent hypotheses tests. The error rate is inflated from 0.05 to 0.226 with five hypothesis tests and to 0.401 with 10 hypothesis tests.

3.6.2 *Frequentist Approach to Multiplicity Problems*

Multiple Testing Formulation

Let H_i $(i = 1, ..., K)$ be the null hypotheses of interest in an experiment. The most common multiple testing problem can be expressed as

$$H_o : \text{all } H_i \text{ are true, versus } H_a : \text{at least one } H_a \text{ is false.} \qquad (3.7)$$

If any H_i is rejected, the global null hypothesis H_o is rejected. For this union-intersection testing, if the global testing has a size of α (also called the familywise error rate or FWER), then this value has to be adjusted to a smaller value when testing each individual H_i. The adjusted value is called the local significance level.

Example: Dose-Funding Trial

In a typical dose-finding clinical trial, where patients are randomly assigned to one of several (K) parallel dose levels or a placebo, the goal is to discover if there is a drug effect and which dose or doses have the effect. In such a trial, H_i will represent the null hypothesis that the ith dose level has no effect in comparison with the placebo. The goal of the dose-finding trial can be formulated in the terms of hypothesis testing (3.7).

The FWER is the maximum (supremum) probability of falsely rejecting H_o under all possible configurations of null hypotheses (e.g., H_1, H_2, H_3, and H_4 are true, or H_1 and H_4 are true but not H_2 and H_3). The strong FWER control at the α level requires FWER $\leq \alpha$.

Single-Step Procedures

Commonly used single-stage procedures include the Sidak method (Sidak 1967) and the simple Bonferroni method. In a single-step procedure, to control the FWER the unadjusted p-values are compared against the adjusted alpha to make the decision as to reject or not to reject the corresponding null hypothesis. Alternatively, we can use the adjusted p-values to compare against the original α for decision-making.

The Sidak method is derived from the simple fact that the probability of rejecting at least one null hypothesis is equal to 1-P(all null hypotheses are retained). To control the FWER, the adjusted alpha α_i for null hypothesis H_i $(i = 1, ..., K)$ can be found by solving the equation

$$\alpha = 1 - (1 - \alpha_i)^K .$$

Therefore, the adjusted alpha is given by

$$\alpha_i = 1 - (1 - \alpha)^{1/K} \geq \frac{\alpha}{K}.$$

Based on this, the simple Bonferroni method conservatively uses the adjusted p-value $\tilde{p}_i = \frac{p_i}{K}$. That is to say, if $\tilde{p}_i \leq \alpha$, we reject the corresponding hypothesis H_i.

Stepwise Procedures

Stepwise procedures are different from single-step procedures in the sense that a stepwise procedure usually follows a specific order to test each hypothesis. In general, such procedures are more powerful than the single-step variety. There are mainly three categories of stepwise procedures: stepup (including the fixed sequence test), stepdown, and gatekeeper procedures (Chang 2011; Dmitrienko, Tamhane, and Bretz, 2010).

Most stepwise procedures can be derived from the so-called closed testing principle or the partition principle. Suppose there are K hypotheses

$H_1, \ldots, H_K$ to be tested at the FWER α. The closed testing principle allows the rejection of any one of these elementary hypotheses, say H_i, if all possible intersection hypotheses involving H_i can be rejected by using valid local-level α tests.

Parallel to the closed testing principle, the partition principle allows for test procedures that are formed by partitioning the parameter space into disjoint partitions with some logical ordering. Tests of the hypotheses are carried out sequentially at different partition steps. The process of testing stops upon failure to reject a given null hypothesis for predetermined partition steps.

Interpretive Paradox of Stagewise Testing

Remember that all stepwise testing procedures can be expressed in a single-step fashion using adjusted p-values. In general, for any given data and any stepwise test procedure, the rejection criterion associated with the null hypothesis H_{ok} at the kth step can be written as

$$p_k \leq \alpha_k(\alpha, p_1, p_2, \ldots, p_{k-1}), \quad k = 1, 2, \ldots K,$$

where the critical point $\alpha_k(\alpha, p_1, p_2, \ldots, p_{k-1})$ is a function of the overall alpha and the previous p-values in the procedure. Equivalently, the above inequality can be written as (by solving for α)

$$\tilde{p}_k(p_1, p_2, \ldots, p_k) \leq \alpha, \quad k = 1, 2, \ldots K.$$

Clearly, the adjusted p-value $\tilde{p}_k(p_1, p_2, \ldots, p_k)$ is a function of all the p-values in the previous and current steps. This implies, paradoxically, that the rejection of a hypothesis H_k will not only depend on the data relevant to the hypothesis, but also on all data collected for other hypotheses $H_1, \ldots, H_{k-1}$ that themselves may have been completely irrelevant to H_k.

False Discovery Rate

The False Discovery Rate (FDR) is the expected proportion of false rejections among all rejections. The FDR is defined to be zero when the total number of rejections is zero (Benjamini and Hochberg, 1995; Westfall et al., 1999, p.21):

$$\text{FDR} = E\left(\frac{\text{Number of rejected true null hypotheses}}{\text{Number of rejected hypotheses}}\right).$$

However, be aware that in the FDR procedure one can purposely include some hypotheses that are known to be rejected in the set of hypotheses to be tested, thus increasing the test level for the hypothesis of main interest (Finner and Roter, 2001).

Suppose we want to test H_o: The drug is ineffective against H_a: The drug is effective, with type I and type II error rates controlled at levels α and β, respectively. If the experiment is repeated, given P_o, the proportion of ineffective test compounds, we have

$$\text{FDR} = \frac{P_o \alpha}{P_o \alpha + (1 - P_o)(1 - \beta)}.$$

Thus, the FDR is a monotone increasing function of α. Furthermore, when $P_o = 1$, the FDR is 100% and so is the type I error rate. In other words, if all test drugs in the pivotal clinical trials are ineffective then, regardless of the α value, 100% of the approved drugs are no more effective than the control. On the other hand, if all drugs in the pivotal trials are effective ($P_o = 0$) then, regardless of α, 100% of the approved drugs are effective (FDR = 0). Test procedures for the FDR can be found elsewhere (Chang, 2011).

Paradox of P-Value in Sequential Experiments

It is often desirable for a cost-sensitive or time-sensitive experiment to have optional stopping in its entire course. For instance, in drug development the effectiveness of a drug is often analyzed several times as patient enrollment progresses and data accumulates. If the test drug is very promising the sequential design will allow the experiment (clinical trial) to stop early, making the drug available to patients sooner while cutting cutting down on costs and resources required. In such a setting, the same null hypothesis H_o (the drug is not effective) is tested several times over the course of the trial. The rejection rules at the ith analysis are usually given as follows:

(1) If $p_i \leq \alpha_i$, stop the trial and reject H_o;
(2) If $\alpha_i < p_{ii} \leq \beta_i$, continue the trial (recruit more patients);
(3) If $p_i \geq \beta_i$, stop the trial and accept H_o.

Here, the constant stopping boundaries α_i and β_i ($i = 1, 2, ..., K$) are so determined that the FWER is controlled (Chang, 2007).

One important question is: In reporting a p-value from a sequential trial, should the p-value be adjusted based on the stage at which the trial stopped (in the spirit of conditionality) or is when the trial stops irrelevant?

3.6.3 Sequential Probability Ratio Test

When a new drug is approved for marketing, pharmacovigilance (a drug safety monitoring system) is critical to ensure patients' safety. A commonly used method for pharmacovigilance is the Sequential Likelihood Ratio Test, also known as the Sequential Probability Ratio Test (SPRT).

Wald's SPRT tests a null hypothesis H_o versus an alternative hypothesis H_a, based on independent observations $x_1, ..., x_i, ...$. As the ith observation arrives, the log-likelihood ratio statistic R_i is calculated as

$$R_i = \ln \frac{l(x_i|H_a)}{l(x_i|H_o)},$$

where x_i represents the cumulative observations from x_1 to x_i.
The stopping rules are as follows:

(1) If $R_i \leq A$, stop monitoring and accept H_o.
(2) If $A < R_i < B$, continue monitoring.
(3) If $R_i \geq B$, stop monitoring and accept H_a.

The stopping boundaries are $A \approx \ln\left(\frac{\beta}{1-\alpha}\right)$ and $B \approx \ln\left(\frac{1-\beta}{\alpha}\right)$, where α and β are the thresholds for type I and type II error rates, respectively. This approximation is based on the assumption of negligible excess of the likelihood ratio over the threshold when the test ends, which is often the case in practice (Chang, 2011).

The likelihood ratio test cannot control the FWER. The question is: What is the upper bound for such error inflation? The Likelihood Principle implies that we can inspect experimental results over time until a certain condition is reached (e.g., in the Paradox of Binomial or Negative Binomial Distribution, we inspect the number of successes until it reaches 3). However, this could cause contradictory conclusions at two inspection timepoints.

Let's examine a simple example of a point null hypothesis H_o against a point alternative H_a: Suppose H_o asserts $\theta = \theta_o$, and the alternative H_a asserts $\theta = \theta_a$. If one is intent on sampling until the likelihood ratio in favor of H_a exceeds r $(r > 1)$, it can be shown that if H_o is true, the probability is at most $1/r$ that one will succeed in stopping the trials (Kerridge, 1963), i.e.,

$$P(LR > r|H_o) \leq \frac{1}{r}.$$

For instance, if $r = 10$, $P(LR > 10|H_o) \leq 0.1$.

3.6.4 *Bayesian Approach to Multiplicity Problems**

Bayesian Multiplicity and Occam's Principle

We should be aware that Bayesian views on multiplicity are different from these of the frequentist. Therefore, the Bayesian approach to multiplicity is different from the frequentist approach. We must also not confuse the penalty on a more complex model (Occam's Razor) with the multiplicity penalty. When competing hypotheses are equal in other respects, the principle recommends selection of the hypothesis that introduces the fewest assumptions and postulates the fewest entities, while still sufficiently answering the question.

Multiple tests in exchangeable settings are handled by a mixed model, $y_i = wf_0 + (1-w)f_1$, where f_0 and f_1 are distributions under noise ($H_o : \theta = 0$) and signal ($H_a : \theta \neq 0$), respectively. The primary goal is to flag which y_i are signals and which are noise.

Empirical Bayesian Approach

The general empirical Bayesian approach to such a multiplicity problem can be outlined as follows:

(1) Represent the hypothesis testing problem, $H_i : \theta_i = 0$, $i = 1, 2, ..., K$, as a model uncertainty problem: Models M_i, with densities $f_i(\boldsymbol{x}|\theta_i)$ for data $\boldsymbol{x}$, given unknown parameters θ_i.
(2) Specify prior distributions $\pi_i(\theta_i)$ and calculate the marginal likelihoods $m_i(x)$.
(3) Specify the forms of prior probabilities, $P(M_i)$, of models to reflect the multiplicity issues.
(4) Implement Bayesian model averaging, based on

$$P(M_i|\boldsymbol{x}) = \frac{P(M_i)\, m_i(\boldsymbol{x})}{\sum_j P(M_j)\, m_j(\boldsymbol{x})}. \tag{3.8}$$

The quantities of interest are $P(M_i|\boldsymbol{x})$, the probabilities of inclusion of θ_i (equivalent to $\theta_i \neq 0$) in the model, and the distributions of the "signals" if the corresponding null hypothesis is not true.

Example: Empirical Bayes Exchangeable Variable Inclusion

To illustrate this with an example, suppose data X arises from a normal linear regression model, with K possible regressors having associated unknown regression coefficients β_i, $i = 1, ...K$. There are many possible submodels with some of the coefficients $(\beta_1, ..., \beta_K)$. For the submodel M_i with k_i regressors, we can choose the prior probability as

$$P(M_i) = p^{k_i} (1-p)^{K-k_i}, \qquad (3.9)$$

where p_i can be the type II maximum likelihood that maximizes $\sum_j p^{k_j} (1-p)^{K-k_j} m_j(\boldsymbol{x})$ for p between 0 and 1.

From (3.8), we can see that the empirical Bayesian approach does control for multiplicity in that $\hat{p}$ will be small if K is large, due to many β_is that are zero. If model M_j includes $\beta_i \neq 0$ and a posterior $P(M_j|\boldsymbol{x}) > p_c$, then β_i will be included. With this approach the false positive rate (inclusion of $\beta_i = 0$ in any submodel M_j) can be controlled. Numerical examples can be found elsewhere (Scott and Berger 2010).

Scott and Berger point out that in the variable-selection problem, if the null model M_0 (including no β_i) has the strictly largest marginal likelihood among all models, then the type II MLE of p is $\hat{p} = 0$. Similarly, if the full model M_F (including all β_is) has the strictly largest marginal likelihood, then the type II MLE of p is $\hat{p} = 1$. As a consequence, the empirical Bayesian approach here would assign final probability 1 to M_0 whenever it has the largest marginal likelihood, and final probability 1 to M_F whenever it has the largest marginal likelihood. These are clearly very unsatisfactory answers. My suggestion to handle this problem is to add an artificial variable $\boldsymbol{x}_0$ known to have the corresponding parameter $\beta_0 \neq 0$ and another variable $\boldsymbol{x}_{K+1}$ known to have the corresponding parameter $\beta_{K+1} = 0$. With these two artificial variables, we can ignore the posterior null model and full model because we know they are noise. Of course, we can add more variables with known β_i's to improve the model's performance.

A good model should have higher probabilities of including any β_is that are truly not zero and lower probabilities of including any β_is that are truly 0. The separation of these two types of probabilities (the true and false inclusion probabilities) is determined by $m_i(\boldsymbol{x})$. Knowing this property is instructive in selecting the density, $f_i(\boldsymbol{x}|\theta_i)$, and prior $\pi_i(\theta_i)$ for model M_i.

Example: Bayes Exchangeable Variable Inclusion

Each variable, β_i, is independently in the model with unknown inclusion probability p. Instead of taking the p in (3.9) from the type II MLE, we can use Beta prior $p \sim \text{Beta}(p|a,b)$. As a special case, a uniform prior on p, with $a = b = 1$, $P(M_i)$ reduces to

$$P(M_i) = \frac{1}{K+1} \binom{K}{k_i}^{-1}$$

Therefore, the posterior model is

$$P\left(M_i|x\right) \propto \frac{1}{K+1}\binom{K}{k_i}^{-1} m_i\left(\boldsymbol{x}\right). \qquad (3.10)$$

As with the situation for the empirical Bayesian approach, it is clear that $P\left(M_i\right) \to 0$ as $K \to \infty$.

Scott and Berger (2010) showed that the prior odds ratio (smaller/larger) reduces as the number of variables K increases, and if that we use the posterior inclusion criterion $P\left(M_i|x\right) \geq p_c = 0.5$ to include β_i, the false positive rate (inclusion of the noise variable) will be controlled.

It should be pointed out that the equal prior probabilities $P\left(M_i\right) = 2^{-K}$ do not control for multiplicity; this corresponds to a fixed prior inclusion probability $p = 1/2$ for each variable.

Note that the control of multiplicity by Bayesian variable inclusion usually reduces model complexity, but is different from the usual Bayesian Occam's Razor effect that reduces model complexity. The latter operates through the effect of the model priors, $\pi_i\left(\theta\right)$, on $m_i(\boldsymbol{x})$, penalizing models with more parameters, whereas multiplicity correction occurs through the choice of the $P(M_i)$. From (3.10), we can see that the order (in terms of value) of $P\left(M_i|x\right)$ is fully determined by $m_i\left(\boldsymbol{x}\right)$.

Regarding how the choice of the prior should depend only on the test experimental procedure, see the discussion in causal space of Bayesianism, Section 3.4.4.

Subjectivity

Subjectivity is nothing but a pretext for all kinds of abductions, including the particular choice of research subject, experiment design, and the process of model selection. Subjectivity exists in every scientific discipline. On the other hand, subjectivity is a reflection of individual experience. Some of our experiences are directly related to the current study, while many others from our past are indirectly or only remotely related to it. Such relationships of past experience to a current study are unclear or difficult to specify; we call such aggregation of knowledge (evidence) a "subjective prior". Without such a subjective prior, everyone would be and act no more maturely than a newborn baby.

Subjectivity does raise concerns in practice, such as when a decision must be made jointly by different stockholders who have different priors, and the decision made will have varying impacts on them. The statistical criterion for NDA (new drug application) approval is one example. The current regulatory criterion is that the p-value should be less than or equal to $\alpha = 5\%$, which we all know is not a good scientific criterion. However, if we use a Bayesian criterion, which prior should be used? Should the prior be different for different candidate drugs? Is it practically and scientifically sound to get a mutual agreement on the prior before we have the clinical trial results? If not, post-study disputes could lead to a disaster.

3.6.5   *Controversies in Multiplicity*

Family of Errors

Frequentists want to control the familywise type I error rate in a multiple testing problem. However, how should we determine the "family" of errors? Should the family chosen be an experiment (e.g., a clinical trial) even when two completely different hypotheses—a single drug treating two different diseases, two drugs treating the same disease, or two drugs treating two different diseases—are included in the same experiment? Why should we adjust multiplicity when hypotheses are addressed in the same experiment but not do so when they are addressed in two different experiments? Does this discourage people who use a more efficient way (one experiment) to do scientific research? Keep in mind that the concept of "an experiment" is subjective. We can describe two experiments as one mixed experiment, as discussed in Section 3.4.1. We can even consider the whole life of an individual to be an experiment, and all the errors one has made in life the family of errors. The difference in the definition of the family of errors could lead to completely different adjustments of p-values and conclusions. On the other hand, controlling alpha beyond the scope of the (commonly defined) experiment is difficult because we don't know what types of studies and how many of them are to be conducted.

We should be aware that the term, type I error does not necessarily mean the same thing at different times. For instance, the first drug for treating a disease may be tested against a placebo. Newer compounds are each year being tested against better and better drugs in the market. It appears that $\alpha = 2.5\%$ doesn't change, but in fact the bar is set higher and higher as different competitor drugs are used.

As we mentioned earlier, a reduction in FDR will increase the chance of treating patients with the right drugs. Thus it seems reasonable to consider FDR control in place of the type I error control, α. However, one can purposely include some hypotheses known to be rejected in the set of hypotheses to be tested and thereby increase the test level for the hypothesis of main interest.

Furthermore, since not all type I errors have the same impact, why do we control the error rate, and not the impact of the errors? Knowing that we only look at hypothesis tests conducted within a company to control the type

I error, how can we control the false positive rate when different ineffective drugs are tested by different pharmaceutical companies?

A Patient's Dilemma

Andy is a biochemist and Mike is a statistician working for the same pharmaceutical company. They both developed cancer and were hospitalized in the same room on the same day. A doctor was discussing their treatment options: drug A or B. Both have been tested on 500 patients. Drug A prolongs survival by 6 months and drug B prolongs survival by 2 years. Andy preferred drug B without a second thought. Mike proceeded cautiously: "Was there any interim analysis and multiplicity adjustment?" The doctor told him that the raw p-value was 0.04 for A and 0.01 for B. However, after the multiplicity-adjustment, the p-value for B was close to significance ($p = 0.055$); no adjustment was needed for drug A. After listening to the details, Mike chose drug A: He thought B was not even statistically significant. Andy asked Mike: "Why did you pick A? Do you believe the statistical testing procedure will damage the chemical structure of the compound?" Mike replied: "The false positive rate will be very high if you choose drugs this way!" Andy wondered: "Everyone only has one life; we don't have many chances to repeat this! Also, the selection of the type I error rate of 5% is a somewhat arbitrary value. Why did Mike take this literally?" Later, however, Mike learned that there was an interim analysis showing that B was better than the control in the trial. Mike informed Andy of this, and asked him if he wanted to switch the treatment to drug A. What would be your answer if you were Andy?

Here, we should exercise caution when considering two concepts. One is the physical properties of the test compound, which cannot be changed by a statistical procedure. The other is the statistical properties, macro properties such as mean and probability, which depend on the causal space (a collection of the similar "experiments" conducted or unconducted) discussed earlier. Such a macro property is a property in the knowledge space.

TO ADJUST OR NOT TO ADJUST?

A PERSONAL DILEMMA

Relationship Between Prior and Multiplicity

Suppose we want to investigate if human genes have any association with disease. We know that when genes are tested independently at a fixed level of significance the familywise error will increase dramatically, due to the large number of genes being tested. From the Bayesian perspective (also from common sense), if we have tested many genes against many diseases and found there is no association between the genes and diseases, then we may start to doubt if there are any associations between any gene and any disease. The more often genes are found to have such disassociations with diseases, the more we believe such disassociation is real in general. On the other hand, if we find that a gene has an association with a disease, then we might believe that more genes will have such an association. When more genes are found to be associated with diseases, the more we believe that other genes and diseases have a similar association. This is exactly the Bayesian notion of a prior; that is, prior knowledge about problem A (e.g., a parameter of a statistical model) is constituted by the knowledge of similar things to A, thus contributing to the final conclusion about problem A. Further thinking along this line leads to another approach to the multiplicity problem. Specifically, in dealing with multiplicity issues, we can treat all genes as one entity (not differentiating among various genes) when we study genes' effects on the same disease. This is one very simple way to deal with the multiplicity issue. To improve performance, we can separate genes into different categories. For genes in the same category, we do not treat them differently in the multiplicity problem, but for genes not in the same category, their associations to the disease can be considered using a discount factor, $r < 1$. However, the determination of the discount factor is somewhat subjective.

In fact, this "discount factor" approach is commonly used in our daily lives and in research, even though we may not fully realize it every time. For instance, we use prior knowledge to choose a particular topic or method for our research. One may say: "I choose this research topic because I need the funding," but such a view or decision is also dependent on one's prior.

3.7 Chapter Review

We discussed the statistical measures of scientific evidence or statistical inference. Statistical inference is commonly used and a powerful tool—we deal with random phenomenon constantly in our scientific research and data collected almost always involve random errors.

We discussed various fundamental principles in statistical inference and controversies surrounding them. Debates on these principles date back a century or more and have become a heated topic again in the past decade.

Despite the philosophical nature of this topic, we took a direct and unsophisticated approach: using paradoxical examples to magnify the problems of each statistical principle. The scope and the hidden assumptions to which each principle applies become apparent under microscopic examination. Many of the paradoxes are in a numerical form, but the numbers are just used to indicate the direction of the conclusion. Without these numerical examples it is very difficult to reach clarity since the topics are difficult conceptually, even for people who are well-trained in statistics.

There are mainly three statistical paradigms: frequentism, Bayesianism, and likelihoodism. Frequentism is centered on the hypothesis test and emphasizes false positive error control; Bayesianism is primarily based on Bayes's law; and likelihoodism is primary based on the law of likelihood.

We reviewed the definition of a statistical model. Using a negative binomial model, we gave an unbiased point estimate. After reviewing the concept of confidence interval (CI), we discussed a paradoxical example of a CI whose construction was based on a single data point, showing you how a single data point can include information about both centrality and variability. We employed a simple classification example to point up the subjectivity of scientific evidence measures.

We classified hypothesis testing into the domain test and point test. We gave a simple example and walked the reader through the hypothesis test, perhaps clarifying some general confusion. Most interestingly, I constructed a paradoxical example indicating that the p-value is not necessarily the evidence against the null hypothesis that we all thought it was.

The five statistical principles (laws) in statistical paradigms are (1) the conditionality principle, (2) the sufficiency principle, (3) the likelihood principle, (4) the law of likelihood, and (5) Bayes's law. Each of these principles deems truth in one or more paradigms without requiring formal proofs.

Bayesianism (the Bayes law) accepts (1), (2), (3), and (5), but not (4). Frequentism accepts (2), but not (3) or (4). Some scholars believe frequentism also accommodates (1), but we have shown you that it is not necessarily true, as illustrated by the paradoxes in Section 3.2.1. (Using Birham's experiment we further showed that the conjunction of the strong likelihood and conditionality principles leads to the likelihood principle that frequentists deny completely!) Frequentists adopt Bayes's law but use it differently from Bayesians. Bayesians use the Bayes law on the basis of two different sets of data (prior and posterior), and emphasizes the learning from prior to posterior. Bayesians consider a model parameter a latent random variable, whereas frequentists consider that a parameter is an unknown but fixed constant. Frequentists use Bayes's law for two random events but not for parameter estimation.

Likelihoodism is relatively simple. It is based on the law of likelihood. It is intuitive in appearance but is logically difficult to justify. After carefully examining Equation (3.4), we can see that the law of likelihood is not exactly the scientific evidence we are looking for.

All five principles are individually appealing but each of them faces multiple challenges, as the paradoxes in Section 3.2 show. The decision approach is based on a utility or loss function. It seems a very logical choice, but it is also fully of controversies, primarily in determining the utility function, which we have discussed under the section of paradoxes in group decisions.

In Section 3.4 we broadened the discussion of controversies in conducting statistical analyses, including a model fitting paradox, Robbins's paradox for irrelevant data pooling, Simpson's paradox for subgroup analyses, regression to the mean and its paradox, and the paradox of confounding.

We proposed the principle of evidential totality and the concept of causal space to unify different statistical paradigms. We used the paradox of the mixture experiment to illustrate why the principle of evidential totality should be used. We discovered and analyzed various paradoxes resulting from different (implicit or explicit) causal spaces. We illustrated the differences in causal space between different statistical paradigms.

Multiplicity issues almost cause impairment to all statistical paradigms. The issue can be simply stated as saying that, due to random variability of the data from scientific experiments and repetitions of the experiments, we could either have a larger number of false positive discoveries or many false negative findings. Multiplicity issues result in a huge waste of time and resources. To deal with the challenging issues, we introduced frequentist and Bayesian approaches and the sequential likelihood ratio test. Unfortunately, none of them is very effective in terms of reducing the false positive and false negative discovery rates. For stepwise test procedures, we pointed out the difficulties in interpretation of the results. The complications of multiplicity issues are further complicated by the controversies in the determination of family of errors and the Patient's Dilemma.

In this chapter we have pretty comprehensively, and in some depth, covered most controversial issues in statistical inference and statistical measures of scientific evidence. In the next chapter we will broaden our discussion of scientific inference and surrounding issues to general science.

3.8   Exercises

3.1 Elaborate the differences between the three statistical paradigms and the principles in each paradigm.

3.2 Explain the following terms: statistical model, unbiased estimator, and confidence interval.

3.3 What is the meaning of p-value? Is the p-value the probability of the null hypothesis being true or the probability of the alternative hypothesis being wrong?

3.4 Does p-value depend on the null hypothesis or the alternative hypothesis or both? How is the level of significance determined in a hypothesis test?

3.5 P-value is commonly considered as evidence against the null hypothesis; is it always true? Why or why not?

3.6 Why do we say scientific evidence involves a subjective component? Can you give an example?

3.7 Explain the five statistical principles: the conditionality principle, the sufficiency principle, the likelihood principle, the law of likelihood, and the Bayes law. How appealing are they intuitively to you? Can we accept them all? Why or why not?

3.8 Compare the Bayes factor and the maximum likelihood ratio. What are the similarities and differences between the two measures?

3.9 What did Birnhaum try to prove in his experiment? What is the implication of his finding?

3.10 In what conditions do the conclusions regarding a hypothesis test derived from a p-value and Bayes's factor become or are likely to become contradictory? In what conditions will the conclusions likely be the same?

3.11 What are the critical components for the decision approach? What is the fundamental challenge in this approach?

3.12 What is the value of model validation and what can be harmful while using model validation?

3.13 Why can we improve the overall precision of parameter estimation by pooling "irrelevant" data together? Give an example other than Robbins's paradox.

3.14 What is "regression to the mean"? Does the paradox of regression to mean essentially deny the presence of regression to mean?

3.15 Use the paradox of confounding to explain the statement: "Statistics can be made to prove anything — even truth."

3.16 What is the Principle of Evidential Totality?

3.17 What is a causal space and how does it vary from paradigm to paradigm? Are we always explicit about the causal space? Can we be?

3.18 Provide real life examples about how differences in causal space (implicit or explicit) lead to different conclusions (outcome space).

3.19 Provide a short description of the multiplicity issue and the Bonferroni method.

3.20 Is a stepwise test procedure usually more powerful than a single-step test procedure? What is a common problem with a stepwise procedure?

3.21 What is a familywise error? What is a false discovery rate?

3.22 What is the challenging issue with the sequential probability ratio test?

3.23 How does the Bayesian approach handle multiplicity issues? Does it control the false positive rate?

3.24 How do you justify the subjectivity in the Bayesian paradigm?

3.25 In the Patient's Dilemma, if you were one of the patients, which drug would you take and why?

3.26 How do you determine the experimental family (or family of errors) when you dealing with a multiplicity issue?

Chapter 4

Scientific Principles and Inferences

Quick Study Guide

As I mentioned in the preface, studying statistical paradoxes helps us understand how scientific evidence is measured, as well as the surrounding uncertainties and controversies, whereas studying scientific paradoxes helps us truly understand the different paradigms and controversies in statistics, and advances the theory of statistics in the right direction.

We are going to study various principles, ideas, rules, methods, inferences, procedures, thought processes, intuitions, and controversies in scientific research. For some of the topics belonging to scientific philosophy, we will use paradoxes to facilitate our discussion and make the abstract concepts tangible and easy to understand with minimum mathematical involvement.

We will discuss several fundamental and provoking questions in scientific philosophy: the definition of science and the meaning of understanding, theories of truth, and discovery versus invention. We give you a fresh view on the topic of Determinism versus Free Will. You will be surprised how these simple terms that we are so familiar with can cause so much confusion and controversies.

We will address the common issues in scientific research, share the paradoxical stories about scientific research, and outline what the gold standard experiment should be. We will introduce a special kind of experiment that does not require a physical experiment and no observations are needed — these thought experiments and examples include Galileo's Leaning Tower of Pisa for free fall bodies, Maxwell's Demon for a perpetual machine, and the Twin and Grandfather paradoxes regarding time travel.

Logical or axiomatic systems are usually considered a topic of mathematics. However, the paradoxes of the logical system we discussed here are way beyond mathematics. The discoveries of these paradoxes have greatly impacted scientific thinking in general, and even computer science, including artificial intelligence.

Game theory has great applications in economics and social science. We will stimulate our discussion with four different paradoxes in game theory.

4.1   Epistemology

Epistemology is the study of the origin, nature, and limits of human knowledge. Important topics in epistemology include: (1) whether knowledge of any kind is possible, and if so what kind; (2) whether some human knowledge is innate (i.e., present, in some sense, at birth) or whether instead all significant knowledge is acquired through experience; (3) whether knowledge is inherently a mental state; and (4) whether certainty is a form of knowledge.

4.1.1   *Science*

There is no consensus definition of science. The most noticeable debates are between evolutionism and creationism. Scientists believe that creationism does not meet the criteria of science and should thus not be treated on an equal footing as evolution. We can say that there is no such thing as philosophy-free science; there is only science whose philosophical baggage is taken on board without examination (Dennett, 1995). It is agreeable that a scientific theory should/can provide predictions about future events; we often take scientific theories to offer explanations for those that occur regularly or have already occurred.

Scientific realists claim that science aims at truth and that one ought to regard scientific theories as true, approximately true, or likely true. Conversely, a scientific antirealist or instrumentalist argues that science, while aiming to be instrumentally useful, does not aim (or at least does not succeed) at truth, and that we should not regard scientific theories as true (Levin 1984).

Analysis is the activity of breaking an observation or theory down into simpler concepts in order to understand it. According to Albert Einstein, the grand aim of all science is to cover the greatest number of empirical facts by logical deduction from the smallest number of hypotheses or axioms.

According to Karl Popper, any hypothesis that does not make testable predictions is simply not science. A scientific theory is constantly tested against new observations. A new paradigm may be chosen because it does a better job of solving scientific problems than the old one. However, it is not possible for scientists to have tested every incidence of an action and found a reaction.

Revolutionary science (paradigm shift) is the term used by Thomas Kuhn in his influential book *The Structure of Scientific Revolutions* (1962) to describe a change in the basic assumptions, or paradigms, within the ruling theory of science. It is in contrast to his idea of normal science.

A scientific revolution occurs when scientists encounter anomalies that cannot be explained by the universally accepted paradigm within which scientific progress has previously been made. The paradigm, in Kuhn's view, is not simply the current theory, but the entire worldview in which it exists, and all of the implications which come with it. It is based on features of the landscape of knowledge that scientists can identify around them. There are anomalies for all paradigms that are brushed away as acceptable levels of error, or simply ignored and not dealt with. Rather, anomalies have various levels of significance to the practitioners of science at the time. Kuhn puts an enhanced emphasis on individuals involved as scientists, rather than abstracting science into a purely logical or philosophical venture.

4.1.2 *Meaning of Understanding*

You may remember that in the story "The Knowing-Fish Bridge" in Chapter 1 we discussed the issue of knowing. What is the meaning of understanding, or the meaning of meaning? If an artificial intelligent (computer) agent can generate "$1 + 1 = 2$" or "It is going to rain tomorrow," do you think she (the agent) understands what she said? Do we know what, if anything, she meant by the sentence? If she generated a text about evolutionary theory before Darwin's and we interpret it as if we are reading Darwin's book, can we say that the agent discovered evolutionary theory? If an agent can randomly or nonrandomly generate any text, then any "invention or discovery" in writing is possible. Similarly, among the vast number of human beings, only a single person, Darwin, founded the science of evolution. Maybe there are others, but we didn't interpret their text correctly. Having said that, it seems that, at end of the day, an invention or discovery depends upon how the readers interpret (or understand) it, not upon how it was written.

Things by themselves are absent meaning. We project meaning onto things depending on our interpretation of their attributes and our current relationship with them. Take the example of a tree. To a farmer clearing his field, a tree would represent rubbish. If you were hiding behind one as someone was firing arrows at you, it would represent a shield.

A thing exists only in relation to the observer(s). It depends the observer's sensor ability. A color-blind person's world is different from a normal person's world. However, that does not prove the "real world" is only a color or black-and-white world. Maybe there are "supermen" can see "super-color." Similarly, our world is three-dimensional, but other creatures may inhabit one that is four-dimensional or two-dimensional. However, that does not prove the "real world" is any particular number of dimensions. Color-blind people describe the world by gray levels, others describe it by color. Therefore, the "truth" is not important, and it can only be defined within a species whose members can communicate well. It does not make sense to talk about a single "real world." The perceptive world is a result of interaction between the "world" and the observer. There is only the perceptive world, not the "world." It is more sensible to use "our world," "my world," than "the world." "The world" exists only for the purpose of making communication convenient.

What is nothingness? It is literally beyond words, beyond concepts, and so can not adequately be described on a merely intellectual level. One creates from nothing. If you try to create from something you're just changing something. So in order to create something, you first have to be able to create nothing (Werner Erhard). Nothing can be created out of nothing, and nothing is more difficult than knowing or doing nothing.

4.1.3 *Theories of Truth*

What is truth? What is a proper basis for deciding how words, symbols, ideas and beliefs may properly be considered true, whether by a single person or an entire society? The theories that are widely shared by published scholars include correspondence, deflationary, constructivist, consensus, and pragmatic theories (Blackburn, and Simmons, 1999).

Correspondence theories assume there exists an actual state of affairs and believe that true beliefs and true statements correspond to the actual state of affairs. Correspondence theory practically operates on the assumption that truth is a matter of accurately copying (what was much later called) "objective reality," and then representing it in thoughts, words, and other symbols (Bradley, 1999).

The deflationary theory of truth is a family of theories that all have a common claim: to assert that a statement is true is just to assert the statement itself. For example (redundancy theory), "I smell the scent of violets" has the same content as the sentence "It is true that I smell the scent of violets." So it seems, then, that nothing is added to the thought by ascribing to it the property of truth.

In contrast to correspondence theories, social constructivism does not believe truth reflects any external "transcendent" realities. Constructivism views all of our knowledge as "constructed," and truth is constructed by social processes, and is historically and culturally specific. Perceptions of truth are viewed as contingent on convention, human perception, and social experience. Representations of physical and biological reality, including race, sexuality, and gender, are socially constructed.

THE TRUTH, THE WHOLE TRUTH AND NOTHING BUT THE TRUTH; BUT IN WHAT SENSE?

Consensus theory holds that truth is whatever is agreed upon, or in some versions, might come to be agreed upon, by some specified group. Such a group might include all human beings, or a subset thereof consisting of more than one person.

In pragmatic theory, "Truth is that concordance of an abstract statement with the ideal limit towards which endless investigation would tend to bring scientific belief, which concordance the abstract statement may possess by virtue of the confession of its inaccuracy and one-sidedness, and this confession is an essential ingredient of truth." There are variations of pragmatic theory (stanford.edu).

A logical truth is a statement that is true under all possible interpretations. For example, a proposition such as "If p and q, then p." is considered to be a logical truth because of the meaning of the symbols and words in it and not because of any facts of any particular world.

According to a survey of professional philosophers and others on their philosophical views, which was carried out in November 2009 (taken by 3226 respondents, including 1803 philosophy faculty members and/or PhDs and 829 philosophy graduate students), 44.9% of respondents accept or lean toward correspondence theories and 20.7% accept or lean towards deflationary theories (Ramsey, 1927).

In my view (Interaction Theory of Truth), if there is any so-called "true object," then the "truth" is a perception of the "true object" or, more precisely, truth is the result of interaction between the "true object" and the observer. In this sense, truth is not unique, but relative to the observer; different observers have different truths. For this reason, there is no difference between illusion and truth. An illusion is the true interaction between the observer and "true object" at a particular time and condition, even though this interaction is so different from interactions experience by other people. This view of truth can be further relaxed. We can say that the so-called "true object" does not exist; it is an abstract concept or a virtual knot that ties together all truths/perceptions from all individuals.

Truth is what we can't prove wrong, not just what we can prove correct. Here "proof of A" means that I can make you believe A more than $\bar{A}$ (negation of A). Thus, any theory about truth can also consider something a truth as long as it can't be proved wrong.

To prove what is correct or what can not be proved incorrect, we have to use a certain language or tool of communication. Thus, we use, for example, words to define meaning and make arguments; but those words are then further defined or explained by other words, ... and on and on. We finally stop either when we believe the final set of words are clear enough or we have no time or energy to continue further! However, we may never be able to prove, for example, that the word "circle" means the same thing for everyone, neither can we prove it means different things for different people. Thus both views can be correct. We will discuss this issue later in Brain and Mind Identity Theory.

A statement is a set of meaningless symbols until it is read and makes a particular sense to the reader.

Paradox of Observer-Dependent Truth

If the world is relative to an observer, then what do we mean by "time" after the observer dies or "time" before the observer is born," and how does one explain the concept of infinite time and the concept of one's father's childhood to the observer?

4.1.4 *Discovery or Invention*

Is science a discovery or an invention? Is it in the external world (outside the human mind) and so must be discovered, or does it lie in the mind and is therefore invented?

Some believe that the only contact human beings have with reality is through the impressions that are received by the senses and the mind. The "objective world" is not, so far as human beings will ever able to tell, completely objective.

According to William Byers (2007, p. 343), "Knowing the truth" is a single unity—both an object and an event, objective and subjective. Knowing and truth are not two; they are different perspectives on the same reality. There is no truth without knowing and no knowing without truth. In other words, the truth is not the truth unless it is known. Nevertheless, "truth" and "knowing" are not identical. They form an ambiguous pair—one reality with two frames of reference. We will discuss more about this when reviewing Fitch's Paradox of Knowability.

Paradox of God Existence

A God believer says: If you believe in God existence, you will be able feel God existence. A nonbeliever says: if I can feel God existence, I will believe God existence. They both are right, in the sense we cannot disprove either of them. The world only exists in the form of the interaction between the "objective world" and the observer. Here, the placeholder "God" can be replaced with anything else. Thus, the paradox becomes the two arguments: "If you believe a thing existence, you will be able feel its existence" and "If I can feel its existence, I will believe in its existence."

A related question is: Is scientific theory an interpretation or an explanation of the world? In my view, science is more of interpretation in nature than an explanation.

4.1.5 *Occam's Razor*

William of Occam (or Ockham) (1284—1347) was an English philosopher and theologian. His work on knowledge, logic and scientific inquiry played a major role in the transition from medieval to modern thought. Occam stressed the Aristotelian principle that entities must not be multiplied beyond what is necessary. This principle became known as Occam's Razor or the law of parsimony: The simplest theory that fits the facts of a problem is the one that should be selected. However, Occam's Razor is not considered an irrefutable principle of logic, and certainly not a scientific result. According to Albert Einstein, "The supreme goal of all theory is to make the irreducible basic elements as simple and as few as possible without having to surrender the adequate representation of a single datum of experience."

According to Gouglas Hofstadter (Mitchell, 2009, preface), Reductionism is the most natural thing in the world to grasp. It's simply the belief that "a whole can be understood completely if you understand its parts, and the nature of their 'sum.' " No one in her left brain could reject reductionism. Reductionism has been the dominant approach since the 1600s. René Descartes, one of reductionism's earliest proponents, described his own scientific method thus: "...to divide all the difficulties under examination into as many parts as possible, and as many as are required to solve them in the best way ... to conduct my thoughts in a given order, beginning with the simplest and most easily understood objects, and gradually ascending, as it were step by step, to the knowledge of the most complex."

Now a simpler model is often less precise but more applicable to a larger set of problems. An over-complicated model is called an *overfitting model* in statistics, which may be more precise for certain problems, but applicable to fewer problems. This touches upon the principle known as Occam's Razor.

Occam's Razor can be most obviously seen in Chinese culture. For instance, the concept of Yin Yang in Chinese philosophy is used to describe how seemingly contrary forces are interconnected and interdependent in the natural world, and how they give rise to each other in turn. Opposites thus only exist in relation to each other. The simple concept, coinciding with Occam's Razor, lies at the origins of many branches of classical Chinese science and philosophy, as well as being a primary guideline of traditional Chinese medicine, and a central principle of different Chinese martial arts. Many natural dualities—dark and light, female and male, cold and hot, the earth and the sky, winter and spring, the moon and the sun—are thought of as manifestations of Yin and Yang, respectively.

YIN & YANG

Yin Yang are complementary opposites that interact within a greater whole, as part of a dynamic system. Everything has both Yin and Yang aspects, but either of these aspects may manifest more strongly in particular objects, and may flow over time. The concept of Yin and Yang is often symbolized by the *taijitu* symbol.

The Five Elemental Energies (*wu sing*) represent the tangible activities of Yin and Yang as manifested in the cyclic changes of nature which regulate life on earth. Also known as the Five Movements (*wu yun*), they define the various stages of transformation in the recurring natural cycles of seasonal change, growth and decay, shifting climatic conditions, sounds, flavors, emotions, and human physiology. Each energy is associated with the natural element which most closely resembles its function and character, and from these elements they take their names. Unlike the Western and other systems of five elements, the Chinese system focuses on energy and its transformations, not on form and substance. The elements thus symbolize the activities of the energies with which they are associated (http://lieske.com/5e-intro.htm).

As manifestations of Yin and Yang on earth, the Five Elemental Energies represent various degrees of "fullness" and "emptiness" in the relative balance of Yin and Yang within any particular energy system. An ancient Chinese text explains this principle as follows:

> By the transformation of Yang and its union with Yin, the Five Elemental Energies of Wood, Fire, Earth, Metal, and Water arise, each within its specific nature according to its share of Yin and Yang. These Five Elemental Energies constantly change their sphere of activity, nurturing and counteracting one another so that there is a constancy in the transformation from emptiness to abundance and abundance to emptiness, like a ring without beginning or end. The interaction of these primordial forces brings harmonious change and the cycles of nature run their course ... The Five Elemental Energies combine and recombine in innumerable ways to produce manifest existence. All things contain all Five Elemental Energies in various proportions.

4.1.6   *Analogies and Metaphors in Scientific Research*

Analogy is an inference or an argument from one particular to another particular, as opposed to deduction, induction, and abduction, where at least one of the premises or the conclusion is general. Analogy plays a significant role in problem solving, decision making, perception, memory, creativity, emotion, explanation, and communication. Analogy is the core of cognition. Specific analogical language comprises exemplification, comparisons, metaphors, similes, allegories, and parables, but not metonymy. Analogy is important not only in ordinary language and common sense but also in science, philosophy, and the humanities.

In science, analogies and metaphors are often used interchangeably. Metaphors can be effectively made across different disciplines. For instance, Allesina and Pascual (2009) recently applied the algorithm PageRank in Google's search engine to ecology and found that a species' PageRank was an excellent predictor of the likelihood that the extinction of that species would lead to secondary extinctions within an ecosystem.

Metaphors can be made from different aspects. Take systems biology as an example. For a machine metaphor, a biological objects (such as cell, organ, organism, population) is thought as a machine; for a language metaphor, biological objects (such as genomes, proteins, metabolic pathways) are thought of as texts written in an unknown script over an unknown language; for informational metaphor, information would be a representable (or communicable) essence of such organization. There are also organic system metaphors and metaphors of organism (Konopka, 2007; p. 21—22).

As Holyoak and Thagard (1995, p. 33) pointed out, "Many cognitive scientists agree that people and other animals make predictions by forming mental models, internal structures that represent external reality in at least an approximate way. Analogy takes us a step beyond ordinary mental models. A mental model is a representation of some part of the environment. For example, our knowledge of water provides us with a kind of mental model of how it moves. Similarly, our knowledge of sound provides us with a kind of model of how sound is transmitted through the air. Each of these mental models links an internal representation to external reality. But when we consider the analogy between water waves and sound propagation, we are trying to build an isomorphism between two internal models."

"Implicitly, we are acting as if our model of water waves can be used to modify and improve our model of sound. The final validation of the attempt must be to examine whether by using analogy we can better understand how sound behaves in the external world." They identified the three constraints to make a good analogy:

(1) *Similarity*: The source of the analogy and the target must share some common properties.
(2) *Structure*: Each element of the source domain should correspond to one element of the target domain, and there should be an overall correspondence in structure.
(3) *Purpose*: The creation of analogies is guided by the problem-solver's goals. Analogies are not fixed forever–as new information comes in, they can be modified.

Biomimetics is the study of the structure and function of biological systems as models for the design and engineering of materials and machines. Biomimetics is technology based on analogy across different disciplines. There are collections of recent developments in biomimetics in Bar-Cohen's book (Bar-Cohen, 2006). One of the most famous examples is Velcro, which is the brand name of the first commercially marketed fabric hook-and-loop fastener, invented in 1948 by the Swiss electrical engineer George de Mestral. De Mestral patented Velcro in 1955 and commercialized it in the late 1950s. The idea came to him one day after returning from a hunting trip with his dog in the Alps. He took a close look at burrs (seeds) of burdock that kept sticking to his clothes and his dog's fur. Under a microscope, he noted that there were hundreds of "hooks" that caught on anything with a loop, such as clothing, animal fur, or hair. He immediately realized the possibility of binding two materials in a simple fashion, if he could figure out how to duplicate the hooks and loops. This inspiration from nature or in analogizing to nature's mechanisms, is what we call biomimetics or bionics.

4.1.7 *Paradoxes of Scientific Reasoning*

Omnipotent Solvent

When a scientist claimed his invention of an omnipotent solvent that can dissolve everything in the world, Edison asked, "What will you store it in if it dissolves everything?" Did Edison make a good logical argument? At first thought, we may think that Edison made a very smart logical argument and proved that there is no such omnipotent solvent. However, anyone who has a little knowledge of chemistry would have a different answer, because there are many possible ways to store the solvent before we use it, for instance, storing at low temperature or in the dark.

Flamingo's Legs

Proximity is viewed as a characteristic of science. A science teacher asked his young students in his class: "How many legs does a flamingo have?" John replied, "Four," while Lisa's answer was "three." The teacher gave a reward to Lisa and said: "A flamingo has two legs, and Lisa gave a better approximation." John, after being silent for a second, said, "Four is the same as two because they both are even numbers, but three is not." Who should get the reward? This is just a naive and humorous version of the "controversy of scientific evidence."

FLAMINGO'S LEGS

Flea's Hearing

Here is another interesting story. A biologist was conducting an experiment to study fleas' behaviors. He was holding a flea in his hand and commanded it to jump; the flea indeed jumped. Then the biologist cut the flea's legs off and commanded it to jump again; the flea didn't jump. Therefore, he concluded, a flea becomes deaf after it loses its legs. We know that there are many possible logical interpretations of a phenomenon, but there is only one "correct" explanation.

Jumping Flea in a Jar

We know fleas can jump extremely high, several times higher than a jar. A scientist put a flea in a jar and 30 minutes later he observed the following:

(1) the height of the flea's jump with the lid on, a little below the lid,
(2) the height of the flea's jump when he removed the lid, a little below the lid's position,
(3) the height of the flea's jump when he put the lid back on, a little below the lid.

Because a flea always jump a little below the lid with or without the lid, he concludes a lid does not contribute to the height of the flea's jump. Here, even a pure statistical analysis on the experimental data can fail.

We now study another version of the same experiment and the conclusion we can draw from it.

We put a flea in a jar and covered it with a lid. For a while the flea continued to try to jump out of the jar, and we could hear the flea hitting the lid of the jar. We watched it until the noise gradually stopped. In fact, the flea continued to jump, but it only jumped a little below the lid to avoid hitting its head. Then, the lid was taken off the jar so the flea could freely jump out of the jar. Surprisingly, the flea continued to jump as if the lid were still on. From these observations, we concluded that the flea learned a lesson quickly from the pain of hitting the lid: It should not jump so high.

Ants That Count!

A story by Robert Krulwich (2009) told of an ingenious experiment conducted in the Sahara suggesting that ants may be able to count. Here, as he gave it, is his story:

How Do Ants Get Home?

Most ants get around by leaving smell trails on the forest floor that show other ants how to get home or to food. They squeeze the glands that cover their bodies; those glands release a scent, and the scents in combination create trails the other ants can follow.

That works in the forest, but it doesn't work in a desert. Deserts are sandy and when the wind blows, smells scatter.

So How Do Desert Ants Find Their Way Home?

It's already known that ants use celestial clues to establish the general direction home, but how do they know exactly the number of steps to take that will lead them right to the entrance of their nest?

Wolf and Whittlinger trained a bunch of ants to walk across a patch of desert to some food. When the ants began eating, the scientists trapped them and divided them into three groups. They left the first group alone. With the second group, they used superglue to attach pre-cut pig bristles to each of their six legs, essentially putting them on stilts.

The third group had their legs cut off just below the "knees," making each of their six legs shorter.

After the meal and the makeover, the ants were released and all of them headed home to the nest while the scientists watched to see what would happen.

The "Pedometer Effect"

The regular ants walked right to the nest and went inside.

The ants on stilts walked right past the nest, stopped and looked around for their home.

The ants on stumps fell short of the nest, stopped, and seemed to be searching for their home.

It turns out that all the ants had walked the same number of steps, but because their gaits had been changed (the stilty ants, like Monty Python creatures, walked with giant steps; the stumpy ants walked in baby steps) they went exactly the distances you'd predict if their brains counted the number of steps out to the food and then reversed direction and counted the same number of steps back. In other words, all the ants counted the same number of steps back!

The experiment results suggest strongly that ants have something like pedometers that do something like counting.

4.1.8 Buridan's Ass: Paradox of Bias

Bias is an inclination to present or hold a partial perspective at the expense of (possibly equally valid) alternatives. Bias can come in many forms. A statistic is biased if it is calculated in such a way that it is systematically different from the population parameter of interest. There are different types of bias.

Selection Bias

Selection bias denotes a selection of subjects into a sample or their allocation to treatment groups that produces a sample not representative of the population, or produces treatment groups that are systematically different. Random selection and random allocation are important ways to avoid this bias.

For instance, a student is conducting an experiment to test a chemical compound's effect on mice. He knows that (placebo) control and randomization are important, so he randomly catches a half of the mice from the cage and administers the test compound. After days of observations, he finds that the test compound has a negative effect on mice. However, he overlooks an important fact: The mice he caught were mostly physically weak and thus they were easily caught. Such selection bias can also be seen in the capture—recapture method due to physical differences in the animals. Another example of selection bias would be regarding an Internet Survey. The bias could be caused by Internet accessibility and how often people use the Internet.

Confirmation Bias

Confirmation bias is a tendency for people to favor information that confirms their preconceptions or hypotheses regardless of whether the information is true. As a result, people gather evidence and recall information from memory selectively, and interpret it in a biased way.

Confirmation bias is remarkably common. Whenever science meets some ideological barrier, scientists are accused of, at best, self-deception, or, at worst, deliberate fraud. Similarly, publication bias is the tendency of researchers and editors to publish experimental results that are positive, but which may or may not be truly positive.

Randomization and blinding in experiments can reduce bias but cannot completely prevent it, especially when there are a larger number of potential confounding factors. We should know that you can tell all the truth, and nothing but truth, but at the same time what you say can be still very biased. For example, you might only talk about the things on the positive side. In a church or other religious society, the same stories are told again and again; these stories often only fit into the religion's needs.

Buridan's Ass is an illustration of a paradox in philosophy in the concept of Free Will. The paradox is named after the 14th century French philosopher

Jean Buridan. It also teaches us not to over-emphasize unbiasedness. The story, briefly, is this: A hungry, thirsty donkey is sitting exactly between two piles of hay with a bucket of water next to each pile, but he has no reason to go to one side rather than the other. So he sits there and dies.

4.1.9 Scientific Method and Experiment

There is no doubt that the most important instrument in research must always be the mind of man (Beveridge, 1957). Having said that, scientific methods also play a critical role in the advancement of sciences.

Scientific method refers to a body of techniques for investigating phenomena, acquiring new knowledge, or correcting and integrating previous knowledge. To be termed scientific, a method of inquiry must be based on gathering observable, empirical, and measurable evidence subject to specific principles of reasoning. Scientific method is a method of procedure consisting of systematic observation, measurement, and experiment, and the formulation, testing, and modification of hypotheses.

The common steps of the scientific method include (1) formulating a question, (2) doing background research, (3) constructing a hypothesis, (4) testing the hypothesis by doing an experiment and collecting the data, and (5) analyzing the data and drawing a conclusion.

The method of scientific investigation is nothing but the expression of the working mode of the human mind. It is simply the mode in which all natural phenomena are analyzed.

Blinded, Randomized, Controlled Experiment

Randomized experiments first appeared in psychology and in education and popularized in agriculture and clinical trials due to two celebrated statisticians, Jerzy Neyman and Ronald A. Fisher, in the 1990s.

A clinical trial is an experiment in human beings to evaluate the efficacy and safety of a candidate drug. Thus, it has the highest ethical and scientific standards. There are three important concepts in clinical trials that generally fit most (if not all) scientific experiments: randomization, control, and blinding. They are the most commonly used, effective ways to control confounders (see Paradox of Confounding in Chapter 3) and reduce bias in the design and conduct of a trial. The bias can be further reduced using proper statistical analyses to control the confounding (e.g., analysis of covariance) as a postexperimental approach. The combination of these four comprises the gold standard for a good experiment.

Control

A control or control group is commonly used in a randomized trial, in which a control (or placebo) is used as a reference group for evaluating the pure effect by the treatment or drug candidate.

A placebo is a simulated or otherwise medically ineffectual treatment for a disease intended to deceive the recipient. The placebo effect is related to the perception and expectation that the patient has: If the substance is viewed as helpful it can heal, but if it is viewed as harmful it can cause negative effects.

It is not unusual that patients given a placebo treatment will have a perceived or actual improvement in a medical condition, a phenomenon commonly called the *placebo effect.* By subtracting the effect in the placebo group from the test group, we can tell the "pure" effect of the drug candidate.

Randomization

The utilization of randomization in a clinical trial is to minimize allocation bias, balancing both known and unknown confounding factors, in the assignment of treatments (e.g., placebo and drug candidate). A proper randomization can also increase the power or efficiency of the trial.

The simplest randomization is called *complete randomization,* in which patients who enter in the trial are assigned, according to a predefined probability, to a treatment group (e.g., placebo or drug candidate).

Blinding

The term "blinded" or "masked" in a trial refers to the procedures that prevent study participants and/or outcome assessors from knowing which treatment was received.

Everyone has opinions and preconceptions. When these opinions are unsubstantiated by evidence, individuals who are involved in a trial can cause bias. In a trial, knowledge of which treatment a patient is assigned to can lead to subjective and judgmental bias by the patients and investigators (Wang and Bakhai, 2006). This bias can influence the reporting and evaluation of the trial outcome, and can even affect the conduct of the rest of the trial. For instance, in a clinical trial patients and doctors can over-report the drug effectiveness due to a psychological effect or to financial incentive.

Statistical Analysis

We are not going to discuss particular statistical methods since one can find them elsewhere easily, though we have discussed numerous controversies in Chapter 3.

4.1.10   *Thought Experiment*

Although experiment without theory is blind, theory without experiment is empty. Here we discuss a special type of experiments, thought experiment. A thought experiment is a mental exercise or virtual experiment. Given the structure of the experiment, it may or may not be possible to actually perform it. The common goal of a thought experiment is to explore the potential consequences of the principle in question. Thought experiments have been used in philosophy, physics, cognitive psychology, political science, economics, social psychology, law, marketing, and epidemiology In fact, many paradoxes are raised from thought experiments, which are used in this book.

LEANING TOWER OF PISA

The well-known thought experiments include Schrödinger's cat, illustrating quantum indeterminacy, and Maxwell's demon, in which a supernatural being is instructed to attempt to violate the second law of thermodynamics. Thought experiments are particularly useful when particular physical experiments are impossible to conduct, such as Einstein's thought experiment of chasing a light beam, leading to Special Relativity. This is a unique use of a scientific thought experiment, in that it was never carried out, but led to a successful theory, proven by other empirical means.

Galileo's Leaning Tower of Pisa experiment was used for rebuttal of Aristotelian gravity. Galileo showed that all bodies fall at the same speed with a brilliant thought experiment that started by destroying the then-reigning Aristotelian account (James R. Brown, 2000). The latter holds that heavy bodies fall faster than light ones ($H > L$). But consider this: a heavy cannon ball (H) and light musket ball (L) are attached together to form a compound object ($H + L$); the latter must fall faster than the cannon ball alone. Yet the compound object must also fall slower, since the light part will act as a drag on the heavy part. Now we have a contradiction. ($H + L > H$ and $H > H + L$). That's the end of Aristotle's theory. But there is a bonus, since the right account is now obvious: They all fall at the same speed ($H = L = H + L$).

Thought experiments can be used to (1) challenge a prevailing status quo or confirm a prevailing theory, (2) establish a new theory, (3) extrapolate beyond the boundaries of already established fact, (4) predict and forecast the (otherwise) indefinite and unknowable future, and (5) explain the past.

There are many different kinds of thought experiments. All thought experiments, however, employ a methodology that is a priori, rather than empirical, in that they do not proceed by observation or physical experiment.

Thought experiments are popular in physics, for instance, Galileo's ship in classical relativity principle; the Bucket argument for arguing that space is absolute, not relational; the monkey and the hunter story to illustrate gravitation, moving magnet and conductor problems; Newton's cannonball in Newton's laws of motion; the Brownian ratchet and Maxwell's demon in thermodynamics; and Einstein's box for the mass—energy relationship. In relativity theory, the well-known thought experiments include, the Ladder Paradox, Twin paradox, Bell's Spaceship Paradox in Special Relativity, and Sticky Bead Argument in General Relativity. For quantum mechanics, the popular thought experiments include the double-slit experiment, Elitzur-Vaidman bomb-tester, GHZ experiment, Heisenberg's microscope, Popper's experiment, quantum pseudo telepathy, and quantum suicide.

No doubt, thought experiments play a significant role in scientific advancement.

4.2   Controversies in Scientific Philosophy

4.2.1   *Determinism Versus Free Will*

If you believe in God you will feel his existence; if you don't believe in God, you will not be able to feel his existence. We cannot prove or disprove such a statement.

Determinism is a broad term with a variety of approaches such as causal determinism, logical determinism, and biological determinisms.

Causal determinism is the thesis that future events are necessitated by past and present events combined with the laws of nature. Imagine an entity that knows all facts about the past and the present, and knows all natural laws that govern the universe. Such an entity may be able to use this knowledge to foresee the future, down to every detail.

Logical determinism is the notion that all propositions, whether about the past, present or future, are either true or false. The problem of Free Will, in this context, is how choices can be free, given that what one does in the future is already determined as true or false in the present.

Biological determinism is the idea that all behavior, belief, and desire are fixed by our genetic endowment. There are other theses on Determinism, including cultural determinism and psychological determinism.

Free Will is the purported ability of agents to make choices free from constraints.

Questions about Free Will and Determinism have been debated for thousands of years. It is a profound problem, because many of us believe that without Free Will there can be no morality, no right and wrong, no good and evil. Were all our behavior predetermined we would have no creativity or choice. However, even some believers in Free Will agree that in every case our freedom is limited by physical reality and the laws of nature, and therefore Free Will is limited.

Albert Einstein said: "I do not believe in freedom of will. Schopenhauer's words, 'Man can indeed do what he wants, but he cannot want what he wants,' accompany me in all life situations and console me in my dealings with people, even those that are really painful to me. This recognition of the unfreedom of the will protects me from taking myself and my fellow men too seriously as acting and judging individuals and losing good humour." (Albert Einstein in Mein Glaubensbekenntnis, August 1932).

"Every event is the effect of antecedent events, and these in turn are caused by events antecedent to them, and so on ... Human actions are no exception to this rule; for, although the causes of some of them are much less well understood than the causes of certain other types of events, it can hardly be denied that they do have causes and that these causes determine their effects with the same certainty and inevitability that are found in every other kind of case. In particular, those human actions usually called free are, each of them, the ultimate and inevitable effects of events occurring long before the agent was born and over which he obviously had no control; and since he could not prevent the existence of the causes, it is clear he could not avoid the occurrence of the effects. Consequently, despite appearances to the contrary, human actions are no more free than the motions of the tides, or the rusting of a piece of iron that is exposed to water and air." (Mates, 1981, p. 59).

Searle (1984, p. 87), on the other hand, believes: If there is any fact of experience we are all familiar with, it's the simple fact that our own choices, decisions, reasoning and cogitations seem to make a difference to our actual behavior. There are all sorts of experiences that we have in life where it seems just a fact of our experience that though we did one thing, we feel we know perfectly well that we could have done something else.

From a neuroscience perspective, it has become possible to study the living brain, and researchers can now watch the brain's decision-making processes at work. A seminal experiment in this field was conducted by Benjamin Libet in the 1980s (Libet et al., 1983), in which it was found that the unconscious brain activity of the readiness potential leading up to subjects' movements began approximately half a second before the subject was aware of a conscious intention to move.

We may believe that identical twins having identical experiences will have identical behavior. In other words, a person's behavior is fully determined by biological composition and the sequence of experiences he had. If that is true, a criminal, on one hand, may say: "I did what I did because of my biological composition and the experiences I've had. Therefore, I should not be punished." The judge, on the other hand, may reply: "The reason I put you in jail is also because my biological composition and experiences." A Christian, who believes in God, may say to the criminal: "You have Free Will and you have a choice not to commit a criminal act." The criminal may offer the defense: "God gave the same biological body to me and my twin brother, but he gave a different 'Free Will' to him." Thus, there are no moral issues without Free Will. This naturally revolves around Richard Dawkins' paradox about Determinism (Gazzaniga, 2011, p. 7).

Free Will implies that a different choice will lead to a different consequence, which is in fact a confirmation of the causality relationship: Everything has a cause. If that is true, how could there be a Free Will?

To put it simply:

According to Determinism,

$$\text{Outcome} = \text{Genetic effect} + \text{Environmental effect}$$

According to Free Will,

$$\text{Outcome} = \text{Genetic effect} + \text{Environmental effect} \\ + \text{Random error due to free-will.}$$

However, if Free Will is a random factor, why do we try so hard to explain the reason for one's action (good or bad)?

We should be aware that the term "identical" can mean the same static structure or the same in both static and dynamic features. For instance, if identical twins' behaviors (output) differ under the same environment (input), then "identical" is not really identical, because even for two machines that look the same, if the same input produce different outputs, we cannot say the two machines are identical. However, what if "identical" means two objects can equally generate unpredictable events, such as sequences of random numbers?

4.2.2 *Paradox of Causality*

Cause and effect is the observation of a repetitive conjunction of particular events that induces us to postulate their linkage. In this philosophical sense, causality is a belief, not a fact. There is a single, growing complex history. To simplify it, people try to identify the patterns or "repetitive conjunction of particular events" and deduce "laws" or causal relationships from the process. The term "cause" relies on two things: the recurrence of the events and definition of "same" or "identical." The recurrence here is in regard to the recurrence of the "cause" and the "effect," both together and separately. However, there is no such exact recurrence. Reoccurrence of an event is only an approximation in the real world; we ignore certain distinctive details, and therefore it is always somewhat subjective.

The notion behind prediction is causality, whereas the notion of causality is the similarity principle: Similar situations will likely result in the same outcome. When we say "the same," we ignore the differences; some of those differences can be hidden and contradict the scientific law to hold. When this happens, an exception to the law is observed. Both science and statistics acknowledge the existence of exceptions, but the former doesn't act until an exception occurs, while the latter acts at the time of the acknowledgment.

Now imagine that you are driving a car; you will know an accident is inevitable a millisecond before it happens. If we anticipate events a millisecond before the "millisecond," by backward induction, we will know things two milliseconds before the accident happens .... Therefore, everything can be determined since God created the universe.

If there is causality, given all the factors that will affect the outcome, the outcome will be uniquely determined or predicted by those factors. Thus "probability" only exists because we lack knowledge about those factors. If we think everything happens for a reason, then the universe is deterministic. If we can "predict" yesterday based on the day before yesterday, we can conclude everything is predictable, at least in theory. However, causality has another requirement: that there are repeated events.

We may have the deterministic view of things at a certain level, but a random perspective at the next detailed level. However, when knowledge increases at the random level, things becomes deterministic while things at the next detailed level becomes random because we don't know enough at this level yet) ... Such a process can repeat over and over.

4.2.3   *Chicken or Egg*

The chicken or the egg paradox is commonly stated as "which came first, the chicken or the egg?" This question also evoked a more general question of how life and the universe began.

The Theory of Evolution answers the question as follows: Species change over time via mutation and selection. Since DNA can be modified only before birth, a mutation must have taken place at conception or within an egg so that an animal similar to a chicken, but not a chicken, laid the first chicken egg. Thus, both the egg and the chicken evolved simultaneously from birds that were not chickens and did not lay chicken eggs but gradually became more and more like chickens over time. The modern chicken was believed to have descended from another closely related species of birds, the junglefowl.

THE CHICKEN OR THE EGG

According to Darwin (Darwin, 1859), " ... if variations useful to any organic being do occur, assuredly individuals thus characterized will have the best chance of being preserved in the struggle for life; and from the strong principle of inheritance they will tend to produce offspring similarly characterized. This principle of preservation, I have called, for the sake of brevity, Natural Selection."

Darwin implied here the four essential conditions for the occurrence of evolution by natural selection:

(1) Reproduction of individuals in the population

(2) Heredity in reproduction

(3) Variation that affects the individual survivals

(4) Finite resources causing competition

The first three are the core requirements (Godfrey-Smith, 2007), the last one, I believe, is not necessary, but it will speed up the evolution process. Those factors result in natural selection that changes the characteristics of the population with an increasing fitness over time.

Evolution theory faces many challenges. Here is an interesting one: Alice, Bob, and Carole are having a shootout. On each round, the current shooter is allowed one shot, aiming at any one of the other shooters. At the start of the game, they draw straws to see who goes first, second, and third, and they take turns repeatedly in that order. A player who is hit is eliminated. Alice is a perfect shot, Bob has 80% accuracy, and Carole has 50% accuracy. Paradoxically, Carole has a 52.2% chance of surviving, whereas Alice has only a 30% chance, and Bob has a 17.8% chance—the survivability is inverse to the accuracy (Gints, 2009; Shubik, 1954).

Darwin's evolutionary theory asserts that natural selection will lead to optimization, but the natural selection algorithm is not an optimized algorithm. There is an apparent difference between the optimal algorithm and the optimal algorithm for finding the optimal algorithm.

4.2.4 Maxwell's Demon

The second law of thermodynamics is an expression of the tendency that over time, differences in temperature, pressure, and chemical potential equilibrate in an isolated physical system. From a state of thermodynamic equilibrium, the law implies the principle of the increase of entropy and explains the phenomenon of irreversibility in nature. In the figure below there are two compartments: The left one is filled with a certain gas, the right one contained the same gas, but at lower temperature. When the sliding door in the middle is opened, the temperature of the gas in the two compartments will gradually become the same, since the fast moving particles in the left compartment will move to the right compartment and slow moving particles will travel from the right compartment to the left one. Such diffusion can be found anywhere. For instance, the "open door policy" between China and the United States will substantially eliminate the (economic) differences between the two countries.

Maxwell's Demon is a thought experiment conceived by the mathematician and physicist James Clerk Maxwell about 150 years ago. Its purpose was to contradict the second law of thermodynamics that declares the impossibility of machines that generate usable energy from the abundant internal energy of nature by perpetual motion of the second kind. (*Less useful* energy means more randomness and more entropy.) In the thought experiment, a small demon operates a friction-free trap door to separate molecules of one type from the other, and finally produces a system with lower entropy. Such an organizing entity of Maxwell's seems to violate the second law of thermodynamics, as the demon only selects molecules but does no work. This paradox kept physicists in suspense for half a century until Leo Szilard showed that the demon's stunt really isn't free of charge. By selecting a molecule he creates the precious commodity called information; there is actually produced an amount of entropy (through mental processing in the brain or any other complicated chemical reaction in a biological system) exactly offsetting the decrement in the rearrangement. The unit of this commodity is the bit, and each time the demon chooses a molecule to shuffle, he shells out one bit ("select" or "un-select") of information for this cognitive act (Mitchell, 2009, p. 43—47).

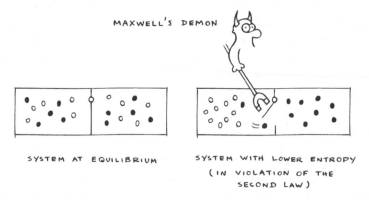

The Brownian ratchet, or Feynman—Smoluchowski ratchet, is a thought experiment about an apparent perpetual motion machine.

First analyzed in 1912 by Polish physicist Marian Smoluchowski and popularized by American Nobel laureate physicist Richard Feynman, the simple machine, consisting of a tiny paddle wheel and a ratchet, appears to be a realizable Maxwell's demon.

4.2.5 *Paradox of Speed and Space*

From classical mechanics, we know that when two cars travel in opposite directions with speeds c_1 and c_2, the speed of either one relative to the other car is the sum of the two speeds $c_1 + c_2$.

Scientists have proved the limit of speed using experiments. However, here is a logical argument that speed has a limit without performing any experiment. Assume there is no limit to speed. A spaceship travels in a straight line. It doubles its speed after half a minute, doubles it again after another quarter of a minute, and continues successively to double it after repeatedly halved intervals of time. Where is it at one minute? It is neither infinitely far away, because there is no such place, nor at any finite distance either, because the distance is $s = 1 + 2\left(\frac{1}{2}\right) + 4\left(\frac{1}{4}\right) + ... + 2n(\frac{1}{2n}) + ... = \infty$. Therefore, there is a limit to speed.

There is a classic argument about whether the Universe is finite or infinite: If it is finite, then when a person throws out a spear at the edge of the Universe, where goes it? Later, there was similar reasoning regarding the finite Universe, noting that if the Universe were infinite with no center (but with a uniform mass density), then a simple calculus can show that universal gravitation would be infinite and pull everything to pieces. In modern physics, the concepts of time, space, and gravity are built on Einstein's relativity theory.

4.2.6 *Modern Conception of Time and Space*

Time and space were viewed as independent dimensions before Einstein conceived of relativity theory. In relativity theory, distances in space or in time separately are not invariant with respect to Lorentz coordinate transformations, but distances in Minkowski spacetime along spacetime intervals are invariant.

In 1905, Albert Einstein published a paper on his Special Theory of Relativity, proposing that space and time be combined into a single construct known as *spacetime*. In this theory, the speed of light in a vacuum is the same for all observers, which has the result that two events that appear simultaneous to one particular observer will not be simultaneous to another observer if the observers are moving with respect to one another. Moreover, an observer will perceive a moving clock to tick more slowly than one that is stationary with respect to her; objects are measured to be shortened in the direction that they are moving with respect to the observer.

Ten years later, Einstein developed the Theory of General Relativity, describing how gravity interacts with spacetime. Instead of viewing gravity as a force field acting in spacetime, Einstein suggested that it modifies the geometric structure of spacetime itself. According to this theory, time passes more slowly at a place of lower gravitational potential, and rays of light bend in the presence of a gravitational field. Scientists have studied the behavior of binary pulsars, confirming the predictions of Einstein's theories. Non-Euclidean geometry (as discussed earlier) is usually applied to describe spacetime.

The Big Bang model or theory is the prevailing cosmological theory of the early development of the universe. According to the Big Bang model, the universe was originally in an extremely hot and dense state that expanded rapidly. This expansion caused the universe to cool and resulted in the present diluted state that continues to expand today.

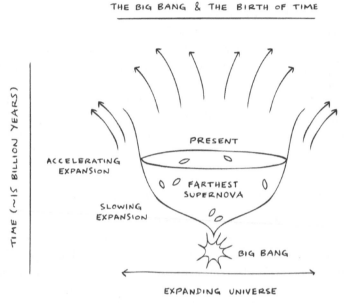

THE BIG BANG & THE BIRTH OF TIME

Extrapolation of the expansion of the Universe backwards in time using general relativity yields an infinite density and temperature at a finite time in the past. This singularity, signaling the breakdown of general relativity, is called "the Big Bang," which can be considered the birth of our Universe, or time zero.

Paradox of Time Partition

According to the theory of relativity, the time between any two events depends on the frame of reference in which the events are measured, i.e., the observer. In terms of possible causality between two events, time can be partitioned into *past, present, future,* and *elsewhere. Past* consists of those events that can influence the present. *Future* consists of events that the present can influence. *Elsewhere* consists of those events that are neither in the past nor the future because they are so far away from the observer that they cannot influence each other, even if their occurrence is transmitted with the speed of light. To illustrate the concept of Elsewhere, since it takes 11 minutes for a signal (at the speed of light) to reach Earth from Mars, an event that takes place on Mars has a 22 minute window of Elsewhere. It's not in the past because it hasn't happened yet for us; it's not in the future because it already did happen and events that are in our present have no ability to influence it. And it isn't in our present because we cannot detect that anything has happened.

Paradox of Continuum of Time

Because time exists only when it associates events (and events associated with observers) and the brain has finitely many states, the human's concept of time should be practically discrete and bounded. If it does not include the time axis, the state-to-reality mapping is not unique. When it includes time, then the brain states form a unique history (a sequence of states). Everything is predetermined. There is only a single string of histories. Everything, including the brain and its thinking, is determined (though we may not know how). We used to separate "thinking" and the "objective world", calling the former "spirit" and the latter "material." But brain and mind (thinking) can be identical; and to separate the world and mind is making mapping between two objective things (mind and world). There is no freewill given by God, or at least not free from God's view. There is no such a thing called causality, and logical reasoning is a fundamental fallency in this sense.

4.2.7 *Time Travel: The Twin and Grandfather Paradoxes*

Time travel is the concept of moving between different points in time, either sending objects backwards in time to some moment before the present, or sending objects forward from the present to the future, without the need to experience the intervening period.

One-way travel into the future is arguably possible given the phenomenon of time dilation, based on velocity in the Theory of Special Relativity, or as gravitational time dilation as in the Theory of General Relativity. It is currently unknown whether the laws of physics would allow backward time travel.

The Twin Paradox

The Twin Paradox is a well-known conundrum from Einstein's Special Theory of Relativity. Take two twins, born on the same date on Earth. One, Albert, leaves home for a trip around the Universe at very high speeds, while the other, Henrik, stays at home, at rest. From relativity theory we know that for an object traveling at high speed, the clock slows down. The question is, when Albert returns home, who is younger? The paradox arises from the following analysis. From Henrik's point of view (and from the viewpoint of everyone else on Earth), Albert seems to speed off for a long time, linger around, and then return. Thus he should be the younger one. But from Albert's point of view, it's Henrik and the whole of the Earth that are traveling, not he. According to special relativity, if Henrik is moving relative to Albert, then Albert should measure his clock as ticking slower, and thus Henrik is the one who should be younger. But this is not what happens. Special relativity predicts that when Albert returns, he will find himself much younger than Henrik.

So what's wrong with our analysis? The key point here is that the symmetry was broken. Albert did something that Henrik did not: Albert accelerated in turning around. Henrik did no accelerating, as he and all the other people on Earth can attest to (neglecting acceleration due to gravity). So Albert broke the symmetry, and when he returns he is the younger one.

The Grandfather Paradox

The Grandfather Paradox is another paradox about time travel: Suppose a man traveled back in time and killed his biological grandfather before the latter met the traveler's grandmother. As a result, one of the traveler's parents would never have been conceived, neither the traveler. This would imply that he could not have traveled back in time after all, which further means the grandfather would still be alive, and the traveler would have been conceived allowing him to travel back in time and kill his grandfather. Thus, each possibility seems to imply its own negation, a type of logical paradox. An equivalent paradox is known as autoinfanticide, going back in time and killing oneself as a baby.

The Grandfather Paradox has been used to argue that backward time travel must be impossible. However, a number of hypotheses have been postulated to avoid the paradox, such as the idea that the past is unchangeable, so the grandfather must have already survived the attempted killing. Some interpretations of time travel also suggest that an attempt to travel backward in time might take one to a parallel universe whose history would begin to diverge from the traveler's original history after the moment the traveler arrived in the past.

Because time exists only as an associated sequence of events (and events associated with observers) and the brain has only finitely many states, the human's conception of time should be discrete, too. Remember, a sequence of events dose not necessarily imply a temporal relationship. When we talk about brain states, the discussion includes time and space and electrochemical properties.

4.2.8 *Concept of Infinity*

In Section 1.3, Mathematical Paradox, we discussed some interesting properties of infinities. Here, we discuss further this important topic.

Infinity is a concept in many fields, most predominantly mathematics and physics, that refers to a quantity without bound or end. In mathematics and philosophy we find two concepts of infinity: potential infinity, which is the infinity of a process which never stops, and static (actual) infinity which is supposed to be static and completed, so that it can be thought of as an object. I embrace the concept of potential infinity; in fact, it is hard to see how we can avoid it.

For the notion of static infinity we have no place in our system of concepts. On the intuitive level, we cannot imagine anything that would qualify as an actual infinity, because neither we nor our evolutionary predecessors ever met anything similar in our experience. When we try to imagine something infinite, for example, infinite space, we actually imagine a process of moving from point to point without any end in sight. This is potential, not actual, infinity (Turchin, 1991). If there is a static infinity, then we have a paradox called Place for Place: Everything has a place for it, thus every place has a place for it, and so on and on....

Kant shows that there is no concept of infinity, only the notion of a rule of a regressus in infinitum. Indeed, Kant and Cantor opened the garden of infinity, where their infinities are only the smallest of bigger infinities.

In my view, the concept of "infinity" can be viewed as the conjunction of the two concepts: "finity" and "more", that is, infinity = more than finity. However, by this definition we have already mapped ∞ to finity, that is, mapping it to the phrases "more than finity." The human brain has limited storage, but it can handle all concepts, including "infinity." If this is truth, then infinity is encompassed by finity, in the brain. So here is the paradox: Infinity is more than finity but at the same time it can be handled by "finity."

There are many examples addressing the relationship between infinity and finity. For instance, $0.\bar{9} = 0.99999999.... = 1$. This is because if $S = 0.\bar{9}$ and $10S = 9.\bar{9}$, so that $10S - S = 9.99... - 0.999 = 9$, and hence $S = 9/9 = 1$. Similarly, we have $1/2 + 1/4 + 1/8 + ... = 1$.

A slightly more complicated example would be the transcendental number (not algebraic—that is, it is not a root of a non-constant polynomial equation with rational coefficients) $\pi = 3.1415...$ It can be computed using a finite-length computer program ("computable"), similarly the transcendental numbers e, $\ln(2)$, e^{π}, Chaitin's constant, and irrational number $\sqrt{2}$. But there are numbers that are incomputable.

To understand the concept of infinity better, let's examine how to mathematically prove the existence of infinity.

Euclid's proof of the infinity of primes can be outlined as follows (Aigner and Ziegler, 2004): For any finite set $\{p_1, ..., p_r\}$ of primes, consider the number $n = p_1 p_2 \cdots p_r + 1$. This n has a prime divisor p. But p is not one of p_i; otherwise p would be a divisor of n and of the product $p_1 p_2 \cdots p_r$, and thus also of the difference $n - p_1 p_2 \cdots p_r = 1$, which is impossible. So a finite set $\{p_1, p_2, \cdots, p_r\}$ cannot be the collection of all prime numbers.

A spaceship travels in a straight line. It doubles its speed after half a minute, doubles it again after another quarter of a minute, and continues successively to double it after repeatedly halved intervals of time. Where is it at one minute? It is neither infinitely far away, because there is no such place, nor at any finite distance either, because the distance is $1 + 2(\frac{1}{2}) + \ldots + 2n(\frac{1}{2n}) = $ infinity. Therefore, there is a limit to speed.

The Paradox of Rare Events

Suppose we randomly select a ball from an urn, where there are infinite number (or billions) of red marbles and one black marble. Before we can find the black ball, the earth is destroyed and human beings disappear. Does it matter if we conclude that there is no black ball? Even if someone has found the black ball, no one else can repeat the "experiment," that is, again find a black ball. Thus, we will conclude that the statement "there is a black ball in the urn" is just an illusion.

4.2.9 *Einstein–Podolsky–Rosen Paradox**

The Einstein–Podolsky–Rosen (EPR) Paradox (Einstein et al., 1935) is a phenomenon in quantum physics and the philosophy of science concerning the measurement and description of microscopic systems by the methods of quantum physics. The EPR paper features a striking case where two quantum systems interact in such a way as to link both their spatial coordinates in a certain direction and their linear momenta (in the same direction). As a result of this "entanglement," determining either position or momentum for one system would fix (respectively) the position or the momentum of the other. EPR use this case to argue that one cannot maintain both an intuitive condition of local action and the completeness of the quantum description by means of the wave function. The authors asked how the second particle can "know" how to have a precisely defined momentum but an uncertain position. Since this implies that one particle is communicating with the other instantaneously across space, that is, faster than light, a "paradox" ensues.

The original EPR paradox challenges the prediction of quantum mechanics that it is impossible to know both the position and the momentum of a quantum particle. This can be extended to other pairs of physical properties. Historically, by 1935 the conceptual understanding of the quantum theory was dominated by Bohr's ideas concerning complementarity. According to Bohr's views at that time, observing a quantum object involves an uncontrollable physical interaction with a classical measuring device that affects both systems. How can we make inferences about past events that we have not observed while at the same time acknowledging that the act of observing affects the reality we are inferring? Thus, the system can only be predicted statistically. The effect experienced by the quantum object restricts those quantities that can be co-measured with precision. According to complementarity, when we observe the position of an object we affect its momentum uncontrollably. Thus, we cannot determine precisely both position and momentum. A similar situation arises for the simultaneous determination of energy and time. This (Heisenberg uncertainty) is also the source of the statistical character of the quantum theory.

Einstein's reservations were twofold. First, he felt the theory had abdicated the historical task of natural science to provide knowledge of significant aspects of nature that were independent of observers or their observations. Instead, the fundamental understanding of the wave function Φ in quantum theory was that it provided probabilities, but only for the outcomes of measurements (the Born Rule). The theory was simply silent about what, if anything, was likely to be true in the absence of observation. Second, the quantum theory was essentially statistical. The probabilities built into the state function were fundamental and, unlike the situation in classical statistical mechanics, they were not understood as arising from ignorance of fine details. In this sense, the theory was indeterministic. Thus, Einstein began to probe how strongly the quantum theory was tied to irrealism and indeterminism (stanford.edu). Einstein asserted the EPR Criterion of Reality: "If, without in any way disturbing a system, we can predict with certainty (i.e., with probability equal to unity) the value of a physical quantity, then there exists an element of reality corresponding to that quantity."

One striking aspect of the difference between classical and quantum physics is that, whereas classical mechanics presupposes that exact simultaneous values can be assigned to all physical quantities, quantum mechanics denies this possibility, the prime example being the position and momentum of a particle. According to quantum mechanics, the more precisely the position (momentum) of a particle is given, the less precisely can one say what its momentum (position) is (stanford.edu). Indeed, the Heisenberg uncertainty principle is understood not just as a prohibition on what is co-measurable, but on what is simultaneously real, a central component in the irrealist interpretation of the wave function.

The debates continues another 15 years, until John Bell constructed a stunning argument, at least as challenging as EPR, but to a different conclusion (Bell 1964). Bell's theorem is often characterized as showing that quantum theory is nonlocal, and it would appear from this theorem that Einstein's strategy of maintaining locality, and thereby concluding that the quantum description is incomplete, may have fixed on the wrong horn of the dilemma. Even though Bell's theorem does not rule out locality conclusively, it would certainly make one wary of assuming it. However, since several other assumptions are needed in any derivation of the Bell inequalities, one should be cautious about singling out locality as necessarily in conflict with the quantum theory.

4.3 Paradox of a Logical System

Logical or axiomatic systems are usually considered a topic of mathematics. However, the paradoxes of the logical system we discussed here are way beyond mathematics. The discoveries of these paradoxes have greatly impacted scientific thinking in general, and even computer science, including artificial intelligence.

4.3.1 *Axiom Systems*

In mathematics, an axiomatic system is any set of axioms, some or all of which can be used in conjunction to logically derive theorems. In an axiomatic system, an axiom is called *independent* if it is not a theorem that can be derived from other axioms in the system. A system will be called independent if each of its underlying axioms is independent. Although independence is not a necessary requirement for a system, consistency is.

An axiomatic system is said to be consistent if it lacks contradiction, that is, the ability to derive both a statement and its negation from the system's axioms. An axiomatic system will be called "complete" if for every statement, either itself or its negation is derivable.

The axiomatic method involves replacing a coherent body of propositions (i.e., a mathematical theory) by a simpler collection of propositions (i.e., axioms). The axioms are designed so that the original body of propositions can be deduced from the axioms.

Non-Euclidean geometry can be viewed as a deductive axiom system with a modification of the parallel postulate, characterized by a non-vanishing Riemann curvature tensor. Examples of non-Euclidean geometries include hyperbolic and elliptic geometry.

The essential difference between Euclidean and non-Euclidean geometry is the nature of parallel lines. Euclid's fifth postulate, the parallel postulate states that, within a two-dimensional plane, for any given line l and a point A, which is not on l, there is exactly one line through A that does not intersect l. In hyperbolic geometry, by contrast, there are infinitely many lines through A not intersecting l, while in elliptic geometry, any line through A intersects l.

Non-euclidean geometry can be understood by picturing the drawing of geometric figures on curved surfaces, for example, the surface of a sphere. Non-Euclidean geometries and in particular elliptic geometry play an important role in relativity theory and the geometry of spacetime.

The concepts applied to certain non-Euclidean planes can only be shown in three or four dimensions. The Möbius strip and Klein bottle are both complete one-sided objects, impossible in a Euclidean plane. The Möbius strip can be shown in three dimensions, but the Klein bottle requires four.

Furthermore, hyperbolic geometry arises in special relativity as follows: An inertial frame of reference is determined by a velocity, and, given a unit of time, each velocity corresponds to a future event from the origin that is the position of an observer with that velocity after the temporal unit. These future events form a hyperboloid, the basis of the hyperboloid model of hyperbolic geometry.

Internal consistency is required in a mathematical axiom system. However, can a person's beliefs be internally consistent throughout their entire lifetime, or for a given time only? Can a person's beliefs change over time? How can we check that a person's beliefs are internally consistent for a given time? After all, checking requires time, and during that time some of his beliefs may already have changed. Furthermore, can a person, even a very intelligent one, truly believe two contradictory theories simultaneously?

4.3.2 *Fitch's Paradox of Knowability**

The paradox of knowability is a logical result suggesting that, necessarily, if all truths are knowable in principle then all truths are in fact known. The contrapositive of the result says, necessarily, if in fact there is an unknown truth, then there is a truth that couldn't possibly be known. Fitch's paradox has been used to argue against versions of anti-realism committed to the thesis that all truths are knowable. It has also been used to draw more general lessons about the limits of human knowledge.

Suppose p is a statement which is an unknown truth; it should be possible to know that "p is an unknown truth." But this isn't possible; as soon as we know "p is an unknown truth," we know p, and thus p is no longer an unknown truth and "p is an unknown truth" becomes a falsity. The statement "p is an unknown truth" cannot be both verified and true at the same time.

The proof of Fitch's paradox can be formalized with modal logic. K and L will stand for known and possible, respectively. Thus, LK means possibly known, in other words, knowable. The modality rules used are (see Section 2.2 Mathematical Reasoning, for the syntax convention; wikipedia.org):

(A) $Kp \rightarrow p$ — knowledge implies truth.

(B) $K(p\&q) \rightarrow (Kp\&Kq)$ - knowing a conjunction implies knowing each conjunct.

(C) $p \rightarrow LKp$ — all truths are knowable.

(D) from $\neg p$, deduce $\neg Lp$ — if p can be proven false without assumptions, then p is impossible (which is the converse of the rule of necessitation: If p can be proven true without assumptions, then p is necessarily true).

The proof proceeds:

(1) Suppose $K(p\&\neg Kp)$

(2) $Kp\&K\neg Kp$ from line 1 by rule (B)

(3) Kp from line 2 by conjunction elimination

(4) $K\neg Kp$ from line 2 by conjunction elimination

(5) $\neg Kp$ from line 4 by rule (A)

(6) $\neg K(p\&\neg Kp)$ from lines 3 and 5 by reductio ad absurdum, discharging assumption 1

(7) $\neg LK(p\&\neg Kp)$ from line 6 by rule (D)

(8) Suppose $p\&\neg Kp$

(9) $LK(p\&\neg Kp)$ from line 8 by rule (C)

(10) $\neg(p\&\neg Kp)$ from lines 7 and 9 by reductio ad absurdum, discharging assumption 8.

(11) $p \rightarrow Kp$ from line 10 by a classical tautology

The last line states that if p is true then it is known. Since nothing else about p was assumed, it means that every truth is known.

4.3.3 Gödel's Incompleteness Theorem*

Kurt Gödel proved in 1931 that any formal system that is complicated enough to deal with arithmetic, that is, contains the counting numbers 0, 1,2,3,4, ... and their properties under addition and multiplication, must either be inconsistent or incomplete. Gödel's incompleteness theorem is important both in mathematical logic and in the philosophy of mathematics. It is one of the great intellectual accomplishments of the twentieth century. Its implications are so far reaching that is difficult to overestimate them.

Gödel's incompleteness proof is very clever. He first invents a brilliant way of coding the elements (symbols, statements, propositions, theorems, formulas, functions, etc.) of the formal system into numbers (called Gödel numbers) and arithmetical operations. Conversely, each such integer can be decoded back into the original piece of formal mathematics. Thus, the proof (disproof) of a statement is equivalent to the proof of the existence (absence) of the corresponding Gödel number. Next, what Gödel did is to construct self-referencing paradoxical statement that says of itself, "I'm unprovable!" in arithmetic. Let's give a sketch of the proof.

Let $G(F)$ denote the Gödel number of the formula (statement, theorem, etc.) F. For every number n and every formula $F(x)$, where x is a variable, we define $R(n, G(F))$ as the statement "n is not the Gödel number of a proof of $F(G(F))$". Here, $F(G(F))$ can be understood as F with its own Gödel number as its argument.

If $R(n, G(F))$ holds for all natural numbers n, then there is no proof of $F(G(F))$. In other words, $\forall x\, R(x, G(F))$, a formula about natural numbers, corresponds to "there is no proof of $F(G(F))$".

We now define the formula $P(y) = \forall x\, R(x, y)$, where y is a variable. The formula $P(y)$ itself has a Gödel number $G(P)$. Replacing y with $G(P)$ in the formula, we have

$$P(G(P)) = \forall x R(x, G(P)). \tag{4.2}$$

This formula concerning the number $G(P)$ is the Liar's paradox: $P(G(P))$ says: "I am not provable." If (4.2) is provable, the left-hand side of (4.2) says $P(G(P))$ is valid, but the right-hand side of (4.2) says "There is no proof of $P(G(P))$. This violates the consistency axiom of the formal theory.

Therefore, (4.2) is unprovable for some formula (function) F that has Gödel numbers as its variables. Thus, the formal system is incomplete.

In conclusion, Gödel's incompleteness theorem proves that a formal system is either inconsistent or incomplete. However, we seem to be more tolerant of incompleteness than inconsistency, and that may be why we have the name Gödel's incompleteness theorem, instead of Gödel's inconsistency theorem. A consistency proof for any sufficiently complex system can be carried out only by means of modes of inference that are not formalized in the system itself. As Gödel stated "Either mathematics is too big for the human mind or the human mind is more than a machine."

However, Gödel's theorems only apply to effectively generated (that is, recursively enumerable) theories. If all true statements about natural numbers are taken as axioms for a theory, then this theory is a consistent, complete extension of Peano arithmetic (called true arithmetic) for which none of Gödel's theorems hold, because this theory is not recursively enumerable. Dan E. Willard (Willard 2001) has studied many weak systems of arithmetic which do not satisfy the hypotheses of the second incompleteness theorem, and which are consistent and capable of proving their own consistency.

4.3.4 Goodstein's Theorem

Gödel's theorem tells us that any formal system has a valid result that is unprovable from the axioms of the system, but it does not give us the actual mathematical statement of a theorem that falls into that category. Goodstein's theorem is a special form of Gödel's incompleteness theorem.

Let's build a series of formulations by choosing an initial number, then performing the following two steps to obtain each consecutive number in the series: (a) Increase the base by 1. (b) Subtract 1 from the number. For example, choosing "45" for the initial number, we have

$$45 = 2^{2^2+1} + 2^{2+1} + 2^2 + 1$$

Step 1 (a): $3^{3^3+1} + 3^{3+1} + 3^3 + 1$
Step 1 (b): $3^{3^3+1} + 3^{3+1} + 3^3$
Step 2 (a): $4^{4^4+1} + 4^{4+1} + 4^4$
Step 2 (b): $4^{4^4+1} + 4^{4+1} + 4^4 - 1 = 4^{4^4+1} + 4^{4+1} + 3 \times 4^3 + 3 \times 4^2 + 3 \times 4 + 3$
Step 3 (a): $5^{5^5+1} + 5^{5+1} + 3 \times 5^3 + 3 \times 5^2 + 3 \times 5 + 3$
Step 3 (b): $5^{5^5+1} + 5^{5+1} + 3 \times 5^3 + 3 \times 5^2 + 3 \times 5 + 2$

Goodstein's theorem proves that the process eventually ends at 0 regardless of the initial number. Goodstein's theorem can be proved (using techniques outside Peano arithmetic) as follows:

Given a Goodstein sequence $G(m)$, we will construct a parallel sequence of ordinal numbers whose elements are not smaller than those in the given sequence. If the elements of the parallel sequence go to 0, the elements of the Goodstein sequence must also go to 0.

To construct the parallel sequence, take the hereditary base-n representation of the $(n-1)^{th}$ element of the Goodstein sequence, and replace every instance of n with the first infinite ordinal number ω. Addition, multiplication and exponentiation of ordinal numbers is well defined, and the resulting ordinal number clearly cannot be smaller than the original element.

The "base-changing" operation of the Goodstein sequence does not change the element of the parallel sequence: replacing all the 4s in $4^{4^4} + 4$ with ω is the same as replacing all the 4s with 5s and then replacing all the 5s with ω. The "subtracting 1" operation, however, corresponds to decreasing the infinite ordinal number in the parallel sequence; for example, ω^{ω^ω} decreases to $\omega^{\omega^\omega} + 4$ if the step above is performed. Because the ordinals are well-ordered, there are no infinite strictly decreasing sequences of ordinals. Thus the parallel sequence must terminate at 0 after a finite number of steps. The Goodstein sequence, which is bounded above by the parallel sequence, must terminate at 0 also.

Goodstein's theorem can be used to construct a total computable function that Peano arithmetic cannot prove to be total—an example of a Gödel incompleteness theorem. The Goodstein sequence of a number can be effectively enumerated by a Turing machine; thus, the function which maps n to the number of steps required for the Goodstein sequence of n to terminate is computable by a particular Turing machine. This machine merely enumerates the Goodstein sequence of n and, when the sequence reaches 0, returns the length of the sequence. Because every Goodstein sequence eventually terminates, this function is total.

4.3.5 *Continuum Hypothesis*

The cardinality of the set of subsets of $\mathbb{N}$ is exactly that of the real numbers, $\mathbb{R}$. The question that arises is whether there are any infinite sets of real numbers that have cardinality strictly greater than that of $\mathbb{N}$ and strictly less than that of $\mathbb{R}$.

Kurt Gödel proved that the continuum hypothesis cannot be disproved using the standard Zermelo—Fraenkel axioms for set theory (including the axiom of choice). Twenty-three years later Paul Cohen showed that the continuum hypotheses cannot be proved from these same axioms either. The hypothesis is independent of these axioms such that there can be two axiom systems with and without the continuum hypothesis. Now the mathematical universe in which the continuum hypothesis is false is richer than the one in which it is true, since it would contain a more diverse variety of subsets of the real numbers.

Does there or does there not exist a set of real numbers with the required intermediate cardinality? In speculating about this question one has the feeling that one is repeating another example of questioning the axiomatic status of a mathematical statement—the great crisis in the history of mathematics that was evoked by the status of another questionable axiom, that is, the parallel postulate of Euclid.

4.3.6   *Paradoxes of Truth and Belief*

Mathematics is irreducible. Logical is indispensable to math and science. Logic moves in one direction, the direction of clarity, coherence, and structure. Is this view agreeable?

There are sometimes no differences between beliefs and truth. For instance, if many people believe Chinese will become a popular international language in the next decade, and so start to learn Chinese, then Chinese will become a popular language. On the other hand, if few people believe Chinese will become a popular language so that almost no one studies it, it will not become a popular language.

A similar example of the coexistence of A and $\neg A$ (the negation of A) would be: Suppose a person, organization, or government is so powerful that his statement will affect future results. Imagine that the chairman of the U.S. Reserve Bank says: "The U.S. economy will not recover within a year." Since his statement will lower consumer's confidence and slow down the U.S. economic recovery, the economy will not recover within a year. On the other hand, if he states: "The U.S. economy will recover within a year," he could also be right, because the statement will increase consumer confidence and speed up the U.S. recovery so that the economy might recover within a year.

Yet another example we discussed earlier is how a patient's recovery from disease sometimes depends on what the doctor informs him. If the doctor says: "You will recover soon," the patient will recover soon due to a psychological effect. On the other hand, if the doctor says: "You will not recover soon," the patient will not recover soon for the same reason. Therefore, the statements A and $\neg A$ are both true—the doctor is always right.

Here are my questions for you: (1) If we don't know how the statement will affect the results, how can we choose between A and $\neg A$, if only one is true? (2) Does the truth of a statement depend on who utters it?

Suppose A is in fact true (happens) and causes B to happen; the statements "$\neg A => B$" and "$\neg A => \neg B$" are always true.

We often say: "You know what you know and you don't know what you don't know." Is this statement true? Is it possible that we're not sure of what we know, or are certain of things we think we don't understand?

4.3.7 *Pigeonhole Principle*

The pigeonhole principle can be simply stated: If you want to put n pigeons into m holes ($0 < m < n$), there will be a hole that has at least two pigeons. The pigeonhole principle is also commonly called Dirichlet's Drawer Principle, in honor of Johann Dirichlet who formalized it in 1834 as "the Drawer Principle."

The pigeonhole principle has many applications. Simple examples would be: (1) Among 13 people there are at least two people having the same birth-month. (2) In any cocktail party with 2 or more people, there must be at least two people who have the same number of friends (assuming friendship is a mutual thing). (3) If some number of people shake hands with one another, there is always a pair of people who will shake hands with the same number of people.

The pigeonhole principle is useful in computer science. For instance, collisions are inevitable in a hash table because the number of possible keys exceeds the number of indices in the array. No hashing algorithm, no matter how clever, can avoid these collisions (wikipedia.org). This principle also proves that any general-purpose lossless compression algorithm that makes at least one input file smaller will make some other input file larger. Otherwise, two files would be compressed to the same smaller file and restoring them would be ambiguous.

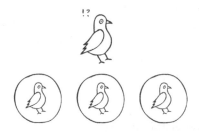

THE PIGEONHOLE PRINCIPLE

A generalized version of the pigeonhole principle states that if n discrete objects are to be allocated to m containers, then at least one container must hold no fewer than $\lceil n/m \rceil$ objects, where $\lceil x \rceil$ is the ceiling function, denoting the smallest integer larger than or equal to x. Similarly, at least one container must hold no more than $\lfloor n/m \rfloor$ objects, where $\lfloor x \rfloor$ is the floor function, denoting the largest integer smaller than or equal to x.

A probabilistic generalization of the pigeonhole principle states that if n pigeons are randomly put into m pigeonholes with uniform probability $1/m$, then at least one pigeonhole will hold more than one pigeon with probability

$$1 - \frac{(m)_n}{m^n}$$

where $(m)_n$ is the falling factorial $m(m-1)(m-2)...(m-n+1)$. For example, if $n = 2$ pigeons are randomly assigned to $m = 4$ pigeonholes, there is a 25% chance that at least one pigeonhole will hold more than one pigeon; for three pigeons and five holes, that probability is 52% (wikipedia.org).

4.3.8   *Paradox of Reductio ad Absurdum*

A common species of reductio ad absurdum is proof by contradiction, where a proposition is proved true by proving that it is impossible for it to be false. Among all principles in philosophy, science, math, engineering, even those in our daily lives, one commonly shared principle is proof by contradiction. It is a cornerstone of all science. However, what if we can use reductio ad absurdum to disprove itself? Here, when I say "Disprove A," I mean to convince you that A is not more believable than $\neg A$.

Reductio Ad Absurdum

Reductio ad absurdum is a mode of argumentation that seeks to establish a contention by deriving an absurdity from its denial, thus arguing that a thesis must be accepted because its rejection would be untenable. In other words, if A being false implies that B must also be false and if it is known that B is true, then A cannot be false, and therefore A is true. It is a style of reasoning that has been employed throughout the history of mathematics and philosophy, from classical antiquity onwards. The principle is almost universally viewed as a mode of argumentation rather than a specific thesis of propositional logic.

The first example is about rational roots of a polynomial equation: There are no rational number solutions to the equation $x^3 + x + 1 = 0$. We prove this by reductio ad absurdum. Assume to the contrary there is a rational number p/q, in reduced form, that satisfies the equation. Then, we have $p^3/q^3 + p/q + 1 = 0$. After multiplying each side of the equation by q^3, the equation becomes $p^3 + pq^2 + q^3 = 0$.

Because p/q is in reduced form, it cannot be the case that both p and q are even. Thus, there are three cases to consider: (1) p and q are both odd, (2) p is even and q is odd, and (3) p is odd and q is even. In all three cases, the left hand side is odd, a contradiction to the right hand side 0. This completes the proof.

The second example is a problem in physical science. We know that matter can transit between phase (gas, liquid, or solid) as temperature changes. For instance, when temperature rises ice becomes water, and water becomes vapor. Interestingly, we can deduce, through proof by contradiction, that there must be a temperature at which both the gas and liquid states of a substance coexist. Assume the lowest temperature for the liquid state of the subtance is C_l and the highest temperature for the solid state of the substance is C_h. If C_l and C_h are different, then there is no such state at temperature $C = (C_l + C_h)/2$. Therefore, C_l and C_h must be the same.

The third example is the thought experiment of Galileo's Leaning Tower discussed earlier.

"Disproof" of Reductio ad Absurdum

We have seen that A and $\bar{A}$ can both be true when the statement itself can influence the outcome. Therefore, in such a case, the proof of contradiction means nothing, which implies reductio ad absurdum is not always valid. A more general disproof of reductio ad absurdum can be given in terms of the Pigeonhole Principle. This is based on the fact that the brain is bounded. Just as Albert Einstein uses the limit of ultimate speed (the speed of light) to derive relativity, here we use a limit of brain states to derive counter-intuitive conclusions.

The number of brain (mind) states (memory) is limited, let's say to 100 billion, then a person cannot completely differentiate numbers 0 to 100 billion because of the pigeonhole principle; this can be proved from Reductio ad absurdum. Of course, memory is also needed to map emotion states. For instance, a feeling for "the infinite" or conceiving of infinity is a state of mind. Therefore, we have the situation where reductio ad absurdum is used to prove the pigeonhole principle, which is used to disprove reductio ad absurdum. We should not confuse the limitation of brain states with the limitation of computer memory. If computer memory is full, we can erase it and put new content in. This is because our brain is larger than a computer, memory-wise. We know what is erased is different from what is put into computer memory later on. But brain state limits will not allow one to differentiate between too many things that are different, so the world will always appear to be limited, even if a person could live forever. For example, a brain state corresponding to a "sweet" feeling will still give the sweet feeling even if one is actually given a salty food, as long as this particular brain state is reached. Another example: If a person has only eight brain states, he will not able to perform a simple single digit addition and subtraction. He may see symbols eight and nine as the same because they correspond to the same brain state.

Because memory and time are limited, we cannot think about thinking about thinking, ..., indefinitely. This self-referential paradox suggests that Aristotle's logic is not conducive to thinking about thinking.

4.4   Paradox and Game Theory

Interestingly, unlike most sciences that deal with the "static" nature, Game Theory deals with games having multiple players, in which one player's actions can affect his opponents' actions. Game Theory has great applications in economics and social science. We introduced four interesting paradoxes. The Prisoner's Paradox presents a case in which individually motivated actions could lead to a globally inferior outcome, i.e., ending with results no one in the group wanted. Braess's Paradox shows a situation in which removing instead of adding a path will reduce the traffic jams in a network. Newcomb's Paradox is created by switching a chronological relation between cause and effect. The paradox of divorce shows in theory that it is always possible to make all marriages stable.

4.4.1   *Prisoner's Paradox*

Prisoner's Paradox is an archetypal example studied in Game Theory. A game is a formal description of a strategic situation. We shouldn't think a game is just a game. Game Theory is the formal study of decision-making wherein several players must make choices that potentially affect the interests of the other players. A player is an agent who makes decisions in a game.

Game Theory is a distinct and interdisciplinary approach to the study of human behavior. Game Theory addresses interactions using the metaphor of a game: In these serious interactions, the individual's choice is essentially a choice of strategy, and the outcome of the interaction depends on the strategies chosen by each of the participants. The significance of Game Theory is dignified by the three Nobel Prizes to researchers whose work is largely Game Theory: in 1994 to Nash, Selten, and Harsanyi, in 2005 to Aumann and Schelling, and in 2007 to Maskin and Myerson.

Most game theories assume three conditions: common knowledge, perfect information, and rationality. A fact is common knowledge if all players know it, and know that they all know it, and so on. The structure of the game is often assumed to be common knowledge among the players. A game has perfect information when at any point in time only one player makes a move and knows all the actions that have been made until then. A player is said to be rational if he seeks to play in a manner that maximizes his own payoff. It is often assumed that the rationality of all players is common knowledge. A payoff is a number, also called the utility, that reflects the desirability of an outcome to a player, for whatever reason. When the outcome is random, payoffs are usually weighted with their associated probabilities to form the so-called expected payoff that incorporates the player's attitude toward risk.

Tucker's Prisoners' Dilemma is one of the most influential examples in economics and the social sciences. It is stated like this: Two criminals, Bill and John, are captured at the scene of their crime. Each has to choose whether or not to confess and implicate the other. If neither man confesses, then both will serve 2 years. If both confess, they will go to prison for 10 years each. However, if one of them confesses and implicates the other, and the other does not confess, the one who has collaborated with the police will go free while the other will go to prison for 20 years on the maximum charge.

The strategies offered in this case are confess or don't confess. The payoffs or penalties are the sentences served. We can express all this in a standard *payoff table* in Game Theory.

The Prisoners' Dilemma

		John (Player B)	
		Confess	Don't
Bill	Confess	$(10, 10)$	$(0, 20)$
(Player A)	Don't	$(20, 0)$	$(2, 2)$

Let's discuss how to solve this game. Assume both prisoners are rational and try to minimize the time they spend in jail. Bill might reason as follows: If John confesses, I will get 20 years if I don't confess and 10 years if I do, so in that case it's best to confess. On the other hand, if John doesn't confess, I will go free if I confess and get 2 years if I don't confess. Therefore, either way, it's better (best) if I confess. John reasons in the same way. Therefore, they both confess and get 10 years. This is the solution or equilibrium for the game. This solution is a noncooperative solution; it is a solution to a noncooperative game in which no communication is allowed and thus no collaboration exists between the players. However, we can see in this example that it is better for both players if they both don't confess, which is a solution of a cooperative game and will be discussed later.

We now introduce an important concept in Game Theory: Nash Equilibrium. Nash Equilibrium, named after John Forbes Nash, is a solution concept of a game involving two or more players in which no player can benefit by changing his strategy unilaterally. If each player has chosen a strategy and no player can benefit by changing his or her strategy while the other players keep theirs unchanged, then the current set of strategy choices and the corresponding payoffs constitute a Nash Equilibrium (see Nash, 1951). The Nash Equilibrium is a pretty simple idea: We have a Nash Equilibrium if each participant chooses the best strategy, given the strategies chosen by other participants. In the Prisoners' Dilemma, the pair of strategies {confess, confess} constitutes the equilibrium of the game.

As I mentioned in Chapter 1, as a paradox related to fixed points, John Nash proved that any zero-sum rational game has a solution (Nash Equilibrium) at which no party in the game will make a move unilaterally.

4.4.2 Braess's Paradox

Braess's Paradox states that eliminating a road, rather than building a road, will sometimes improve traffic conditions.

When 42nd Street in New York City was temporarily closed to traffic, rather than the expected gridlock, traffic flowed more easily. In fact, as was reported in the September 2, 2002, edition of *The New Yorker*, in the 23 American cities that added the most new roads per person during the 1990s, traffic congestion rose by more than 70% (Havil, 2008). Braess's Paradox was named after the German mathematician Dietrich Braess, who published a paper (Braess et al., 1968, 2005) in which he showed that, under the appropriate conditions, building new roads to ease congestion actually makes the problem worse.

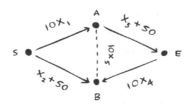

BRAESS' PARADOX

Suppose there are initially two alternative roads to E from S: $S \to A \to E$ and $S \to B \to E$. The one-way relief road $A \to B$ was added with the intention of easing congestion. The time taken on each route is shown in the figure below, where x_i is the number of vehicles on that road and a labels aside a line indicates the travel time. Assume that each driver is completely aware of the traffic situation on these five roads at all times and selfishly chooses the route that minimizes his time of driving.

Suppose there are six cars. Let's look into the process consisting of these six cars, with and without the relief road.

(1) Without the relief road $A \to B$, there will be three cars on route $S \to A \to E$ and three cars on route $S \to B \to E$. The average driving time is $10(3) + 3 + 50 = 83$ mins.

(2) With road $A \to B$, we assume all six cars line up at the starting point (S), one by one. Because each driver takes the shortest route available at the moment, the first two cars will be on $S \to A \to B \to E$ with the drivers' perceived driving time 60 minutes, shorter than the alternative routes. The third and fourth cars will be on $S \to A \to E$ and $S \to B \to E$, with the drivers' perceived driving time 81 minutes, one on each route; the fifth and sixth cars will also be on $S \to A \to E$ and $S \to B \to E$, with the drivers' perceived driving time 92 (minutes). However, the drivers' perceived time is not the actual time because they didn't consider the drivers behind them. The actual drive time should be based on the actual number of cars on the routes: $x_1 = 4$ cars on $S \to A$, $x_2 = 2$ cars on $S \to B$, $x_3 = 2$ cars on $A \to E$, $x_4 = 4$ cars on $B \to E$, and $x_1 = 2$ cars on $A \to B$. The driving time on route $S \to A \to E$ or $S \to B \to E$ is $10(4) + 2 + 50 = 92$. The driving time on route $S \to A \to B \to E$ is $10(4) + 10(2) + 10(4) = 100$. It is interesting that the first two drivers, who think they are getting the best deal for being the first two to make the choice, are actually getting the worst result, the longest driving time.

Braess's Paradox can be found at play in any network, such as water flow, computer data transfer, electrical and electronic networks, and telephone exchanges. In 1990, the British Telecom network suffered in such a way when its "intelligent" exchanges reacted to blocked routes by rerouting calls along "better" paths. This in turn caused later calls to be rerouted with a cascade effect, leading to a catastrophic change in the network's behavior (Havil, 2008).

Little is known about whether Braess's Paradox is a common real-world phenomenon, or a mere theoretical curiosity. If it is a rare event in selfish routing networks, then such strategies might be largely superfluous for real-world networks. On the other hand, if the paradox is a widespread phenomenon, then the problem of adding capacity (or new road) to a selfish routing network must be treated with care. Recently, Valiant and Roughgarden (2010) proved that there is a constant $\rho > 1$ such that, with probability 1 (i.e., $1 - O(n^{-c})$, $c > 0$) as $n \to \infty$, a random n-vertex network admits a choice of traffic rate such that the resulting Braess ratio is at least ρ. The Braess ratio of a network is defined as the largest factor by which the removal of one or more edges (roads) can improve the latency of traffic in an equilibrium flow. Chung and Young (2011) showed that Braess's Paradox occurs when $np \geq c\log(n)$ for some $c > 1$, where p is edge probability. Nagurney (2010) showed that as demand (total traffic) increases, the Braess Paradox "works itself out." One would expect that at a higher level of demand the network gets even more congested and that more of the paths/routes would then be used. However, Nagurney (2010) showed that the route that resulted in the Braess paradox at a particular level of demand will no longer be used at a higher level of demand (zero traffic). This suggests that there may be an underlying "wisdom of crowds phenomenon" taking place. But how far from optimal is traffic at equilibrium? It can be as bad as one can imagine, but under the assumption of a linear "potential energy," it can only be, at worst, twice as bad as socially optimal (Von Ahn, 2008).

4.4.3 Newcomb's Paradox

Newcomb's Paradox, is a thought experiment created by William Newcomb of the University of California in 1960 (Wolpert and Benford, 2010). However, it was first analyzed and published in a philosophy paper spread to the philosophical community by Robert Nozick in 1969, and appeared in Martin Gardner's *Scientific American* column in 1974.

The paradox goes like this: The player of the game is presented with two boxes, one transparent (labeled A) and the other opaque (labeled B). The player is permitted to take the contents of both boxes, or just the opaque box B. Box A contains a visible $1. The contents of box B, however, are determined by an omnipotent being who can predict everything correctly. The Omnipotent Predictor told the player: "I will make a prediction as to whether you will take just box *B*, or both boxes. If I predict both boxes will be taken, I will put nothing in box *B*. If I predict only box *B* will be taken, then I will put $1,000 in Box *B*." At some point before the start of the game, the God-like being made the prediction and put the corresponding contents (either $0 or $1,000) in Box *B*.

The player is aware of all the rules of the game, including the two possible contents of box *B*, the fact that its contents are based on the Predictor's prediction, and knowledge of the Predictor's infallibility. The only information withheld from the player is what prediction the Predictor made, and thus what the contents of box *B* are.

The problem is called a paradox because two strategies (the Dominance Principle and the Expected Utility Maximization), both intuitively logical, give conflicting answers to the question of what choice maximizes the player's payout (wikipeda.org).

Scenario	God-Like Prediction	The Player's Choice	Payout
1	*A* and *B*	*A* and *B*	$1
2	*A* and *B*	*B* only	$0
3	*B* only	*A* and *B*	$1001
4	*B* only	*B* only	$1000

The Dominance Principle: This strategy argues that, regardless of what prediction the Predictor has made, taking both boxes yields more money. That is, if the prediction is for both *A* and *B* to be taken, then the player's decision becomes a matter of choosing between $1 (by taking *A* and *B*) and $0 (by taking just *B*), in which case taking both boxes is obviously preferable. But, even if the prediction is for the player to take only *B*, then taking both boxes yields $1001, and taking only *B* yields only $1000—taking both boxes is still better, regardless of which prediction has been made.

The Expected Utility Maximization: This strategy suggests taking only *B*. By this strategy, we can ignore the possibilities that return $0 and $1001, as they both require that the Predictor has made an incorrect prediction, and the problem states that the Predictor is never wrong. Thus, the choice becomes whether to receive $1 (both boxes) or to receive $1000 (only box *B*)—so taking only box *B* is better.

In his 1969 article, Nozick noted that "To almost everyone, it is perfectly clear and obvious what should be done. The difficulty is that these people seem to divide almost evenly on the problem, with large numbers thinking that the opposing half is just being silly."

The crux of the paradox is in the existence of two contradictory arguments, both being seemingly correct.

If there is no Free Will, I don't have a choice at all. My future actions are determined by the past prediction of God. Since I don't have the Free Will to make choice, the game does not make sense at all. Without Free Will, I even wonder what is the meaning of "if."

If there is Free Will, then I have choice and my choice will affect God's prediction, meaning, my future action will determine the fate of a past event (reverse causation), such as will be described in the Grandfather Paradox later.

Interestingly, Newcomb said that he would just take B; why fight a God-like being? (Wolpert and Benford, 2010).

A similar paradox, called the Idle Argument, goes like this: Suppose that the Omnipotent Predictor has predicted the grade for my future exams. Then I can just relax and sit, doing absolutely nothing at al.

4.4.4 Stable Marriage Problem

Suppose there are n males and n females going to get married. Every male has a ranked list of females he likes, and vice versa. In the matching process, a married couple will choose divorce if they both have a better choice; otherwise, they will stay together. The question is: Given two preference lists for the men and women, is there a stable marriage, meaning no further divorce will occurs? By the way, here we apply U.S. marriage law: A man can have one and only one wife, and vice versa.

Intuitively, we may say "No" since we have seen so many divorces continuously occur in real life. Surprisingly, the answer is positive. The Proposal Algorithm proposed by Gale and Shapley (1962) proved that, for any equal number of men and women, it is always possible to solve the Stable Marriage Problem and make all marriages stable.

The Gale-Shapley algorithm can be described as follows:

Each unengaged man M proposes to his most-preferred woman W to whom he has not yet proposed. If she (W) is free, they are engaged. If W has already engaged with a more preferable man M', then W and M' stay as couple. If W has engaged with a less preferable man M', then W engage with M and M' becomes a free man. Such processes continue until there is no free man.

This algorithm guarantees that everyone gets married, and all the marriages will be stable. This is because if there is a free woman, there must be a free man who will propose. For such a proposal, if the list of engaged couples cannot be further improved, the man will engage with a free woman; otherwise, the list will be modified or improved. Such improvement cannot only be done with limited times, therefore, a free man will eventually engage with someone already engaged or a free woman.

While the solution is stable, it is not necessarily optimal from all individuals' points of view.

Algorithms for finding solutions to the stable marriage problem have applications in a variety of real-world situations, perhaps the best known of these being in the assignment of graduating medical students to their first hospital appointments (Skiena 1990, p. 245).

4.5 Chapter Review

Epistemology aims at answering many fundamental questions about the origin, nature, and limits of human knowledge. There is no single consensus definition of science. Scientific realists claim that science aims at truth and that one ought to regard scientific theories as true, approximately true, or likely true. Conversely, scientific antirealists don't agree. They believe science aims for instrumental usefulness. Along this line, we further discussed the theories of truth and meaning of understanding, including the correspondence, deflationary, social constructivist, the consensus, pragmatic, and the interaction theories of truth. We pointed out that "understanding" is nothing but a word-mapping or concept-mapping process.

We conceptually differentiated the terms: discovery (originated from external world) and invention (originated within a piece of the mind). However, it is so difficult to name any particular subject a discovery or an invention (See also Paradox of Discovery in Chapter 5).

Occam's razor asserts that the simplest theory that fits the facts in a problem is the one that should be selected to explain it. However, what is missing in Occam's razor is the clarity. The conception of Yin Yang follows Occam's razor. Many Chinese consider Ying Yang as the source for every single thing and its dynamics. But its abstractness obviously lacks clarity, also.

We outlined the common steps of the scientific method (research), pointed out the issues of confounding and bias, and emphasized the importance of blinding, randomization, and control in an experiment.

It is true that scientific discovery is often achieved through a well planned experiment and careful follow-up observations. However, there are special kinds of experiments that do not require a physical experiment or any observations. That is, thought experiments. A powerful and efficient tool to discover truth is to carefully set up an imaginary experiment, a thought experiment, and several rigorous steps of logical reasoning. The idea of thought experiments was exemplified by Galileo's Leaning Tower of Pisa experiment.

We emphasized the importance of analogy in scientific research and analyzed an example in Biomimetrics.

We discussed scientific reasoning and controversies using five mini-stories: omnipotent solver, a flea's hearing, flamingo's legs, fleas jumping, and ants that count.

We further discussed controversies in scientific philosophy. In debating Determinism versus Free Will we contended that Determinism will not cause ethical issues since punishing one's action is not a matter of choice. Further along this line, we claimed that the term *causality* only makes sense when there will be reoccurrences of the same event. Such, a recurrence of "cause-effect" pairs is summarized as a law. However, if everything happens for a reason, the path of each thing will be unique. Thus we group similar ("same") things together.

We outlined the necessary conditions for Natural Evolution and listed some successful applications of the notion of Natural Evolution in various fields.

A perpetual motion machine is a hypothetical machine that operates forever by generating more energy than it consumes. The Maxwell demon is invented to challenge the second law of thermodynamics, which states that a perpetual machine is impossible.

We discussed the classic and modern concepts of space, time, and speed in conjunction with the concept of infinity. The paradoxes of speed and space illustrate some classic thought experiments to prove that space and speed are limited. In modern physics, Einstein considered space, time, and gravity together in his general relativity. According the Big Bang theory, time has a starting point and the universe is expanding.

We usually partition time chronologically into past, present, and future. However, physicists find there is another component of time, called "elsewhere." Past consists of those events that can influence the present. Future consists of events that the present can influence. Elsewhere consists of those events that are neither in the past nor the future because they are so far away from the observer that they cannot influence each other, even if their occurrence is transmitted with the speed of light.

Relativity theories postulate that we can slow down aging, and go back to our past, if we travel at a speed close to the speed of light. However, the Twin Paradox and the Grandfather Paradox are two thought experiments that are crafted to challenge this claim.

The concept of infinity is always interesting and a controversial topic to discuss. There are physical infinities (e.g., of time, space, speed) and mathematical infinity (e.g., the infinitely small, infinitely large, and infinite processes). We discussed their relationships. In fact, we cannot clearly separate the two types of infinity. To understand the concept of infinity better, we showed by example how one can prove mathematically the existence of infinities.

The EPR (Einstein—Podolsky—Rosen) paradox has provoked a great debate in quantum physics and the philosophy of science. It remains a true paradox. The paradox is a thought experiment, in which two quantum systems interact in such a way as to link their spatial location and momenta. When quantum particles are thus entangled, it is impossible to know both the position of one and the momentum of the other. This relates to the issue of whether uncertainty in quantum mechanics is essential.

Probably nothing is considered more rigorous than logical or axiomatic systems. However, the paradoxes of logical system have made a great impact on scientific thinking in general. The best known examples are probably non-Euclidean geometries, which allow the sum of three angles of a triangle to be larger or smaller than 180 degrees. This has changed the face of modern physics in dealing with the micro and macro universes.

Fitch's paradox of knowability proves that if all truths are knowable in principle then all truths are in fact known. Therefore, there are only two possible truths, (1) the truths we already know, and (2) the truths that are unknowable.

Gödel's Incompleteness Theorem states that an (complex enough) axiom system is either incomplete or inconsistent. An example of Gödel's Incompleteness Theorem exists in Goodstein's theorem.

The continuum hypothesis postulates that there are no infinite sets of real numbers that have cardinality strictly greater than that of the natural number system and strictly less than that the real (number) system. It has been proved that we cannot either prove or disprove the hypothesis. This leads to a creation of another richer system that does not require the continuum hypothesis.

We often use the law of the excluded middle in proof and disproof. That is, given a statement and its opposite statement, one and only one is correct. But its implicit assumption that the statement will not affect the outcome is neglected. The paradox of truth and belief reveals the hidden assumption.

Probably no scientific law is simpler than the Pigeonhole Principle. It says that if you try to put a certain number of pigeons in a smaller number of holes, one of the holes must have more than one pigeon. The principle has found wide applications. But interestingly, when applying this principle to a human brain with a limited memory, many concepts—even the concept of truth—become strangely unclear. We don't even know that Reductio ad Absurdum should be a general scientific law.

Interestingly, unlike most sciences that deal with "static" nature, the science of Game Theory deals with games having multiple players, in which one player's action will affect his opponent's action. Game Theory has great applications in economics and social science. We introduced the concept of game and stimulated discussions with the four well-known paradoxes. The prisoner's paradox presents a case in which individually motivated actions could lead to a globally inferior outcome, that is, ending with the result that no one in the group wanted. Braess's Paradox shows a situation in which removing instead of adding a path will reduce the traffic jams in a network. Newcomb's Paradox is created by switching a chronological relation between cause and effect. Despite the increased divorce rate in our society, the paradox of divorce shows that, in theory, for any given number of men and women it is always possible to make all marriages stable.

4.6   Exercises

4.1 What are the important topics in epistemology?

4.2 How do you define science?

4.3 What is the meaning of understanding?

4.4 What are the main theories of truth?

4.5 How do you define discovery, invention, and belief? Is mathematics an invention or a discovery? Is religion a discovery, invention, or belief? Why?

4.6 Is Occam's Razor something necessary or just nice to have?

4.7 Please elaborate the importance of analogies in scientific research.

4.8 Can you identify some typical fallacies in scientific research?

4.9 Identify sources of biases in scientific research and methods to reduce biases.

4.10 What is a randomized controlled experiment? What is the purpose of randomization? Why do we use a placebo or control in an experiment?

4.11 Are you able to identify any hidden assumptions in the thought experiment of Galileo's leaning tower?

4.12 Do you believe in the brain—mind identity theory? Why or why not?

4.13 Do you believe in Free Will? If you do, where did your Free Will come from? If it came from somewhere else and you didn't have any "will" to make a decision on whether to accept it or not, who should have the responsibility for your actions in your life? On the other hand, if you don't believe in Free Will, do you agree that a man is just like a machine without emotion? In such a case, what is the meaning of "should" and social morality?

4.14 If the Universe is deterministic, would there be any meaning of causality? If everything is determined by Free Will, would there be such a thing called science?

4.15 Explain the time component "elsewhere."

4.16 Explain the concepts of static infinity and potential infinity with examples.

4.17 What is the hidden assumption in the argument presented in the paradox of infinity?

4.18 Do you think time should be continuous or discrete? Why?

4.19 Do you have a resolution to the Grandfather Paradox? Have you heard of the parallel universe?

4.20 What are the four essential conditions for evolution to occur? Can you identify some applications of the notion of Natural Evolution to other fields?

4.21 Why does Maxwell's Demon fail to prove the possibility of a perpetual machine?

4.22 Write an essay to address the importance of axiom systems and the significance of Gödel's theorem.

4.23 Fitch's paradox of knowability states: If all truths are knowable in principle then all truths are in fact known. Do you believe this? If you do, we don't need to do scientific research to discover truths anymore. If you don't believe it, please disprove the knowabiliy.

4.24 Identify some applications of the pigeonhole principle.

4.25 Why can the limit of brain memory cause a problem in theory and philosophically?

4.26 What is the most fundamental difference between Decision Theory and the Game Theory?

4.27 How can learning the Prisoner's Paradox and Braess's Paradox benefit one's life?

Chapter 5

Artificial Intelligence

Quik Study Guide

This chapter has two distinct parts, the paradox in artificial intelligent (AI) and the architecture for a new type of AI agents. Those who are not interested in AI building or do not have a computer science or AI background may choose to skip the second part.

The idea of AI was proposed by Alan Turing in the Turing Machine. There are great debates on whether any computer can be built to perform like a human. I assert this can simply be a definitional problem, as illustrated by the "Ship of Theseus." The possibility of building "machine-race" humans is discussed through paradoxes of emotion, understanding, and discovery. The possibility is further illustrated by the paradox of dreams, the paradox of meta-cognition, and recent developments in swarm intelligence.

There are several fundamental differences between the proposed AI approach and the existing AI approaches: (1) The new AI has very limited built-in initial concepts (only 14)—virtually an initial empty brain, whereas traditional AI has a built-in knowledge database and/or complex algorithms. (2) The new AI is language-independent, whereas traditional AI is language-dependent (English and Chinese AIs have different architecture). (3) Unlike traditional AI, the new AI is based on the recursion of simple rules over multiple levels in learning. (4) The new AI's learning largely depends on progressive teaching: It can learn very broadly about many things but it is slow, whereas traditional methods have a fast learning speed but a narrow learning scope.

5.1   Paradoxes in Artificial Intelligence

5.1.1   *From Turing Machine to Artificial Intelligence*

The ideas embedded in the Turing machine marked the start of the computer age, and described and established the possibility of artificial intelligence.

A Turing machine is a theoretical device that manipulates symbols on a strip of tape according to a table of rules. Despite its simplicity, a Turing machine can be adapted to simulate the logic of any computer algorithm. The Turing machine was described by Alan Turing in 1937, which was not intended as a practical computing technology, but rather as a thought experiment representing a computing machine.

In his 1948 essay, "Intelligent Machinery," Turing described a Logical Computing Machine, which consisted of "... an infinite memory capacity obtained in the form of an infinite tape marked out into squares, on each of which a symbol could be printed. At any moment there is one symbol in the machine; it is called the scanned symbol. The machine can alter the scanned symbol and its behavior is in part determined by that symbol, but the symbols on the tape elsewhere do not affect the behavior of the machine. However, the tape can be moved back and forth through the machine, this being one of the elementary operations of the machine."

A Turing machine that is able to simulate any other Turing machine is called a Universal Turing machine. A more mathematically oriented definition with a similar "universal" nature was introduced by Alonzo Church, whose thesis states that Turing machines indeed capture the informal notion of an effective method in logic and mathematics, and provide a precise definition of an algorithm or "mechanical procedure."

According to Russell and Norvig (2006), there are mainly four approaches to AI: thinking rationally (the "laws of thought" approach), acting rationally (the rational agent approach), thinking humanly (the cognitive modeling approach), and acting humanly (the Turing test approach).

A rationally thinking computer agent should be able to use logical reasoning. To act rationally, computer agents (rational agents) are expected to act so as to achieve the best outcome. To think like a human, we must have some way of determining how humans think. We need to probe the actual workings of human minds. Cognitive science can provide helpful insights, but there are still many controversies. To act humanly, an agent employs operationally defined intelligence, as defined by the Turing test, proposed by Alan Turing (1950). Rather than proposing a long and perhaps controversial list of qualifications required for intelligence, Turing suggested a test based on indistinguishability from undeniably intelligent entities—human beings. The computer passes the test if a human interrogator, after posing some written questions, cannot tell whether the written responses come from a human or not.

A true artificial intelligent agent should have the following important capabilities:

- It can learn any natural language and eventually carry on conversations with a human in a meaningful way.
- It has various sensors for vision, hearing, smell, taste, and touching.
- It can synergize all information from different sensors in the learning process.
- It can assimilate knowledge and acquire and generate new knowledge.
- It can interact with peer agents physically and intelligently.
- It can communicate human motion by responding like a human.

There are obvious differences between approaches centered around humans and approaches centered around rationality. A human-centered approach must be an empirical science, involving hypothesis and experimental confirmation. A rationalist approach involves a combination of mathematics and engineering.

From an application point of view, AI can mean search algorithms, machine learning, data mining, or an AI agent (AIA). In this loose sense, AI has been studied and used in various disciplines. The AIA to be discussed in this book is the human-like AIA; such an AIA will be able to act rationally because it will be able to learn the things that are needed to act rationally. I call such a human-like AIA a machine-race human (MRH).

5.1.2 *Brain and Mind Identity Theory*

Mind is the term most commonly used to describe the higher functions of the human brain, particularly those of which humans are subjectively conscious, such as personality, thought, reason, memory, intelligence, and emotion. Although other species of animals share some of these mental capacities, the term is usually used only in relation to humans. It is also used in relation to postulated supernatural beings to which human-like qualities are ascribed, as in the expression "the mind of God."

The brain is the part of the central nervous system situated within the skull. It includes two cerebral hemispheres, and its functions include muscle control and coordination, sensory reception and integration, speech production, memory storage, and the elaboration of thought and emotion.

BRAIN-MIND IDENTITY THEORY

The Language of Thought Hypothesis (LOTH) postulates that thought and thinking take place in a mental language. This language consists of a system of representations that is physically realized in the brain of thinkers and has a combinatorial syntax (and semantics) such that operations on representations are causally sensitive only to the syntactic properties of representations. According to LOTH, thought is, roughly, the tokening of a representation that has a syntactic (constituent) structure with an appropriate semantics. Thinking thus consists of syntactic operations defined over such representations. Most of the arguments for LOTH derive their strength from their ability to explain certain empirical phenomena, like productivity and systematicity of thought and thinking (Murat Aydele, cogprint.org, stanford.edu).

Functionalism in the philosophy of mind identifies mental states and processes by means of their causal roles, and we know that the functional roles are possessed by neural states and processes. Thus, it is reasonable to suspect that the way in which the brain represents the world might not be through language. The representation might be much more like a map. A map relates every feature on it to every other feature. We can think of beliefs as expressing the different bits of information that could be extracted from the map.

A functionalist believes, according to the token identity theory, that a particular pain (more exactly, having a pain) is identical to a particular brain process. Functionalism came to be seen as an improvement on identity theory, and as inconsistent with it, because of the correct assertion that a functional state can be realized by quite different brain states: Thus, a functional state might be realized as well by a silicon-based brain as by a carbon-based brain.

5.1.3 *Machine-Race Human*

There have long been debates about whether a human-like machine can be made. Before we can give a clear answer, we have to agree on a definition of what it means to be human. It might come as a complete surprise that the term "human" can be perceived so differently by different people. According to wordiq.com, human beings are defined variously in biological, spiritual, and cultural terms, or in combinations thereof. Biologists classify human beings as *Homo sapiens*, a primate species of mammal with a highly developed brain, belonging to the family of great apes, along with chimpanzees, bonobos, gorillas, and orangutans. Theologists describe human beings in spiritual terms, using various concepts of soul that, in religion, are understood in relation to divine powers or beings; in mythology, they are also often contrasted with other humanoid races. Anthropologists define human beings by their use of language, their organization in complex societies, their development of technology, and especially by their ability to form groups and institutions for mutual support and assistance. Other definitions are more or less along the same lines. However, such a definition is not satisfactory in our current discussion about AI, which can be seen in the following identity paradox.

5.1.4 *The Ship of Theseus: Identity Paradox*

Heraclitus's "river fragments" raise puzzles about identity and persistence: Under what conditions does an object persist through time as one and the same object? If the world contains things which endure, and retain their identity in spite of undergoing alteration, then somehow those things must persist through changes. Can one step into the same river twice precisely if it continually undergoes change (Washington.edu)?

Here is another example. Suppose there is a ship that in the year 1800 sails out to sea. This ship is made of 100 boards, a sail, and a mast, and is called "The Ship of Theseus." A year goes by with the ship out to sea, when a single board of the ship's original 100 is replaced by brand new wood. The ship sails on. Another year goes by with the ship out to sea and another original board of the ship is replaced by brand new wood. After 100 years, in 1900, all 100 original boards have been replaced by 100 new boards. Finally, all that's left to replace is the sail and the mast. And so it happened that in the year 1901, the sail was replaced, and in 1902, the mast. The question is: should the ship have its original identity and be called "The Ship of Theseus," or should it have a completely different identity (www.unc.edu)?

We now discuss how "I" as an individual is defined. Suppose that for medical reasons I have had my heart, liver, kidney, legs, and stomach replaced over the years. Should the replaced person still be called "me" or someone else? What if I had my brain repaired or replaced or my memory erased or copied from someone else? Should such a person be called "me" or someone else or nobody? If two people A and B have exchanged half of their brains and some other body parts, who is A and who is B now? If we can make human organs replace most of a man's organs, then would this person be an artificial man or a classical human being? And how should he be treated? Can we discriminate for or against him? If a person eventually replaces all his organs or his whole body, should this person be called an artificial being instead of a human being? What if we can clone a person? What if we could join a biological human and computer? What if a man-made embryo is injected into a human body and raised within that human body?

When an artificial person is so highly developed that we cannot differentiate it from a "real" human being without knowing how it was made, then our human society will become a union of human and artificial. Humans and the definition of "human" will start to change.

From the discussions above, if we cannot clearly define the entity that is a person, how can we expect to have a consensus on whether AI can be really human or not? Moreover, following the same logic, we cannot clearly or unambiguously define what emotion is; therefore, it is inconclusive as to whether AI can have emotion or not.

5.1.5   *Paradox of Emotion*

According to the interaction theory of the next section, like any other concept, the meanings of "understanding" and "emotion" are particular states of a brain. Dreaming something can give the same sensation (at the time of dreaming) as the corresponding real event. A blind person can "see" (experience sensation) when he is stimulated using an electronic device that connects to his back. Therefore, we don't try to create an agent with emotion or an ability to understand using only conventional wisdom.

Emotion can be a result of frequent interactions with someone or something else. If one constantly plays with a cat, an emotional bonding will be established. Most people experience emotional exchanges with their pets or other animal. Similarly, a person who plays a role character in a game often feels strongly emotions attached with the character, and might say "I am the person in the game." If in the future an AIA behaves like a human and interacts closely with humans, then the definitions of "human" and "emotion" will evolve. If and when this happens, AIAs and classical humans will unite and form a single human society. There will then be one more kind of human being in this society, i.e., the machine-race human.

Modern science tells us that the DNA of any two humans is 99.97% identical. The 0.03% difference in DNA and environmental differences make every one of us a unique person. However, nearly everyone uses and relies on the Internet so much nowadays that it becomes a strong homogenizing force, wiping out many of the difference among us. We may develop virtually the same social norms, and same emotions toward the same events. The Internet turns people all over the world into supermen because they share the same "brain," while popular websites such as google.com, wikipedia.org, etc., serve as different parts of our "brain."

The influence exerted between humans and an AI agent is a two-way interaction. A human creates and teaches or "raises" an AIA. At the same time, the agent can change the human's behavior, just as the Internet does.

5.1.6 *Paradox of Understanding*

Can an agent possess "understanding"? But what do we mean by "understanding?" It relates to the meaning of the term "concept." *Concepts*, pretheoretically, are the constituents of thoughts. There are five common issues regarding concepts: (1) the ontology of concepts, (2) the structure of concepts, (3) empiricism and nativism in regard to concepts, (4) concepts and natural language, and (5) concepts and conceptual analysis. Understanding is judged by the consistency between what is observed and what can be explained through reasoning and past experience. If you are laughing while hearing a joke, it would be judged that you got (understood) it. But if a baby was laughing at that very moment, he/she may not be judged "understanding," because our reasoning and experience tell us that the baby cannot understand the joke.

We also need to clarify a few things before we can begin our discussion. That is, a concept or term is defined by other concepts and terms, which are further defined by other words and terms, and so on. Thus, some people believe a basic initial set of words is necessary in communication. This first set of words is used to explain other words or concepts; more words can then be explained further by those words that are just explained, etc. The meaning of each word is different for different individuals and changes over time. Most of the time people think they have understood each other when they establish a mapping (or agreement) between what they perceive. It is basically the situation: "I know what you are talking about" or "I understand what you mean." What I think your understanding is and what your actual understanding is can be different. But it does not matter as long as I think I understand that you understand, and vice versa.

Imagine that we connect words by arrows to form a network, called a wordsnet. Let's consider the wordsnet of an English dictionary. For example, starting with the word "mother," which is defined as "a female parent," in the Webster's; we use three arrows to connect "mother" to the three words "a," "female," and "parent." Then these three words are further defined by other words, and linked by arrows, ..., and so on. In this way, you are constructing a wordsnet, which turns out to be not an acyclic network as one might expect, but a cyclic one with many loops. Similarly, the wordsnet of a human brain is also an acyclic network, but one with many loops. Furthermore, the conceptsnet (a similar network connecting concepts) of a human brain is a cyclic network with loops. You would see something like this (recursive definition):

confusion means perplexity; perplexity means bewilderment, bewilderment means incomprehension; and incomprehension means confusion.

We see that confusion is actually defined by confusion, but after going through many such loops (often much bigger loops), people feel they are getting an understanding of the meaning of the word or concept.

5.1.7 *Paradox of Discovery*

Can an AIA discover new knowledge? It depends how we define the term "discover." For instance, can an AIA discover "evolution theory"? What does the term "discover" mean here? If an AIA did generate the exact same text as Darwin's evolution theory, does it mean that the theory has been "discovered"? Should a "discovery" be determined by its "discoverer," the AIA, or the reader (the human)? Should we explain the text differently from Darwin's text just because it was generated by a machine? If we do explain the machine-generated text the same way as we did in Darwin's text, should we call it a discovery? If we do, we in fact have created an AIA that can make scientific discoveries because such an AIA is just a random text generator.

When a machine can be made "human" is not only dependent on the advancement of computer (AI) technology, but also dependent on how we change our view (including moral standards) of the concept of "human," and on the extent to which we can accept the machine-race human as human without racial discrimination.

5.1.8 *Paradox of Dreaming*

Everyone dreams every day. Dreams are fascinating. They are characterized by successions of images, ideas, emotions, and sensations that occur involuntarily in the mind during certain stages of sleep. The content and purpose of dreams have been a topic of scientific speculation, philosophical intrigue, and religious interest.

Oneirology, the study of dreams, shows that dreams mainly occur in the rapid-eye movement (REM) stage of sleep, when brain activity is high and resembles that of being awake. REM sleep is revealed by continuous movement of the eyes during sleep. At times, dreams may occur during other stages of sleep. However, these dreams tend to be much less vivid or memorable. Dreams can last for a few seconds, or as long as twenty minutes. A person is more likely to remember the dream if he or she is awakened during the REM phase (wikipedia.org). REM occurs in other mammals as well as in humans, but not in other types of animals. Moreover when people are kept out of REM sleep (by being awakened whenever REM starts), they show a stronger tendency to enter it. Whatever is going on in REM sleep apparently is necessary. According to Winson (1985), a basic problem in biological adaptation is how to integrate new experiences with old ones; such integration takes place during REM sleep.

However, I'd argue that when we speak of a dream it is often not the real dream or what actually happens in one's brain that we refer to. Instead, it is one's recollection of the dream when awakened, which could be completely different from the actual dream. Dreaming of something happening and that thing actually happening might give a person the same sensation, but at different times, one while sleeping and the other at the awakening. We really don't know if our "dream" is truly a dream or what we thought a reality is in fact a dream. In a dream, a reality might be thought a dream and the dream a reality; while when we wake up, we believe a dream is a dream. Therefore, there is an unclosable gap between dreams and what one recalls about them, just as between the worlds of life and death. What we think to be reality might be embedded in a bigger dream.

Despite the controversy, the period of dreaming is a useful time for each of us (or an AIC) to digest and organize what we have learned in the prior "daytime". A dream can also be a creative process.

5.1.9 *Paradox of Meta-Cognition*

The brain certainly defines the range of our neural possibilities, not only what we can do but also what we cannot even imagine doing. But at the same time, one of the hardest problems is trying to understand the mind while using it as a tool. This requires us to use a tool to define a tool or a process to define a process (www.community.middlebury.edu).

Cox and Raji (2011, p. 3) pointed out: "Philosophers and cognitive scientists of many persuasions have long wondered what is unique to human intelligence. Although many ideas have been proposed, a common differentiator appears to be a pervasive capacity for thinking about ourselves in terms of who we are, how others see us, and in terms of where we have been and where we want to go. As humans, we continually think about ourselves and our strengths and weaknesses in order to manage both the private and public worlds within which we exist. But the artificial intelligence (AI) community has not only wondered about these phenomena; it has attempted to implement actual machines that mimic, simulate, and perhaps even replicate this same type of reasoning called metareasoning."

After all, asking the question "Can we think about thinking?" is equivalent to asking "Can one lift oneself up off the ground by pulling one's hair with one's hands?"

Thinking about thinking (TAT) is a type of thinking, and thus we have the concept of thinking about TAT (denoted by TATAT), etc. A simple example of this chain of thinking would be: TAT is to recall the thought process that happened yesterday; TATAT is to recall the recollection process yesterday about the thought process that happened the day before yesterday. We should be aware that thinking about how others think is quite different from thinking about how oneself thinks. The former is imagination.

5.1.10 *Swarm Intelligence*

Swarm intelligence (SI) or collective intelligence (CI) is the collective behavior of decentralized, self-organized systems. In such a system, no individual is in charge or plays the central rule. Each individual in a SI has no concept of collaboration with its peers. Natural examples of SI include ant colonies, bee colonies, bird flocking, animal herding, bacterial growth, and fish schooling.

The most well-known example may be that of ant colonies. An individual ant is pretty dumb, but a colony of ants is fairly intelligent. This perplexed scientists for a long time. How was it that a colony of dumb ants could find food so efficiently without a leader and carry out other functions so intelligently?

As illustrated by Marques de Sa (2006, p. 199—203), a large set of living beings manifest learning skills needed for survival. Ants are able to determine the shortest path leading to a food source. The process unfolds as follows. Several ants leave their nest to forage for food, randomly following different paths. Those that by chance reach a food source along the shortest path sooner reinforce that path with pheromones (a chemical produced by an organism that signals its presence to other members of the same species), because they sooner come back to the nest with food; those that subsequently go out foraging find a higher concentration of pheromones on the shortest path, and therefore have a greater tendency (higher probability) to follow it, even if they did not follow that path before. Such a learning mechanism is called *sociogenetic learning*. The learning curve here resembles the convergence of a "spatial average." The time required for learning is inversely proportional to the number of individuals involved in the activity.

Swarm intelligence can also be artificial. Such an SI system is typically made up of a population of simple agents. The inspiration often comes from nature, especially biological systems. The agents follow very simple rules, and although there is no centralized control structure dictating how individual agents should behave collaboratively, the relatively free interactions between such agents lead to the emergence of intelligent global behavior. Borrowing words from Thomas Schelling, the 2005 Nobel Prize winner for Economic Science, SI is characterized by micro motives and macro behavior (Schelling, 2006).

Interestingly, not long ago Southwest Airlines was wrestling with a difficult question: Should it abandon its long-standing policy of open seating on planes? Of all the major airlines, Southwest was the only one that let passengers choose where to sit once they got on board. The company's independent attitude had helped make it one of the largest airlines in the world. Lately, though, some customers, especially business travelers, had complained that the free-for-all to get on a Southwest plane was no fun. To obtain a good seat, travelers had to arrive at the airport hours before their flight to secure a place at the head of the line. In the competitive airline business, the company was willing to make a change if it made passengers happier. The question is what is the best boarding strategy? Boarding speed depends upon the boarding strategy, for example, start boarding passengers in the back of the plane and work forward, or go in the opposite direction? What about boarding window seats first, then the middle seats, then, finally, aisle seats? It was believed that ant colonies were a good model for the study, because every individual has a task, obtaining a seat, while dealing with all the others doing the same thing (Miller, 2010). After the artificial ant simulation, Southwest figured out the that best strategy is to assign the seats at check-in, but boarding would still be first-come, first-served. Now this strategy has become a standard for various airlines.

About fifteen years ago, Bloom (Waibel, Floreano, and Keller, 2011) showed how the collective intelligences of competing bacterial colonies and human societies can be explained in terms of computer-generated complex adaptive systems. Bloom traced the evolution of collective intelligence to our bacterial ancestors 1 billion years ago, and demonstrated how multispecies intelligence has worked since the beginning of life on our planet (Dorigo and Stützle, 2004).

Everything on earth is made of molecules. In this sense living things are made of nonliving things. Scientifically, individual genetics and environmental factors uniquely determine one's behavior: Two biologically identical persons (having identical Free Will, if you are a believer in God) who are placed in the same environment will behave identically. In this respect, a person is just like a machine.

Collective intelligence is closely related to *complexity*. An ant is simple, while a colony of ants is complex; neurons are simple, but brains are complex. Competition and collaboration between cells leads to a human intelligence; competition and collaboration between humans form a social intelligence, or what we might call the global brain. Nevertheless, such intelligence is based on a human viewpoint, and thus it lies within the limits of human intelligence. Views of such intelligence held by other creatures with a different level of intelligence could be completely different!

5.1.11   *Man in World in Man*

Mankind is part of nature. We often talk about the relationship between man and nature or the environment. But by doing this, humans actually put themselves above nature. A man lives in the world; that is "man in the world." A man has a perception about the world; that is "world in man." Therefore, we have a second-level recursion: "Man in World in Man (MWM)." Similarly, there can be even higher levels of recursion between a man and the world.

There are potentially two distinguishable AI approaches: (1) to view the human subject and nature as one entity, (2) to separate the human subject from nature; then what we discuss is nature's reflection in ourselves, not necessarily nature itself—if the latter has a definition. This second approach is more popular than the first one. The perception of the world in the second approach is believed to be resident in the human brain, but then different people have different worlds or realities.

The world is relative to the observer. In other words, what the world appears to be will depend on the abilities of the observer's eyes, nose, ears, ..., and communication abilities. Therefore, the "truth" of the world or of a worldly event can only be defined within a specie. Indeed, communication is probably the most fundamental aspect when defining the categories of living species. Communication may also be used to differentiate "living" and "non-living" species. The former group evinces communication among peers, but the latter does not. What is communication then? Communication is the mapping process between mindstates of two individuals (pairwise communication).

Our concepts of the external world (outside the brain/mind) associate the states of the brain. For the purpose of illustration, let's code different brain states using binary numbers. For the same thing (concept), the corresponding brain states for two different people, e.g., one English speaker and one Chinese speaker, can be different. However, mappings between any two approximately identical concepts are established through direct or indirect communication.

BRAIN-WORLD MAPPING

BRAIN STATE	CONCEPT IN AN AMERICAN	CONCEPT IN A CHINESE	BRAIN STATE
0000	CONCEPT OF CONCEPT —— 概念之概念		0000
0001	1+1=2 ——————— 1+1=2		0001
0010	TRUTH	无限	0010
0011	INFINITE	真实的	0011
0100	OPPOSITE ——————— 相反		0100
0101	FALSE	矛盾	0101
0110	FREE WILL	虚无	0110
0111	CONTRADICTORY	错误的	0111
1000	LOVE ——————— 爱		1000
1001	TIME ——————— 时间		1001
1010	EMPTY	反证法	1010
1011	REDUCTIO AD ABSURDUM		1011

Most of us have had the experience that some words in our language do not translate exactly into another. This is often because people from different countries/regions have different concepts about certain things or group them differently. For the very same reason, we cannot expect the same word in a language to describe exactly the same concept for two different people, even if they speak the same language.

5.1.12   *Crafting a Child or an Adult Agent*

Although we have achieved significant progress in AI, an important reason why we have not achieved greater success, in my view, is that we have focused on what a human can do, but completely ignored what a human can't do. "Can" and "cannot" are equally important in building a truly human-like AI agent. When we try to construct a weak aspect of our agent, we are automatically building a miraculous part of the AI agent. Why is that?

In the past, we tended to create AI agents by implementing complicated theories in mathematics, statistics, and decision or game theory, including many methods in datamining, textmining, genetic programming, and other disciplines. We focus on the agent's ability to answer our questions and provide us with solutions (or entertainment). We judge an agent based on the ability to solve some specific challenges or tasks. I'd like to call this approach the one-time-investment approach or single-stage approach because when you have built the agent only a minimal investment is needed in his future learning process. The agent will be able to perform tasks or learn from the data or environment automatically. Such an agent has limited abilities and will only be able to carry out some particular tasks, and its learning abilities are very limited.   This is just like a man who, when he gets old, loses his curiosity, creativity, is less open-minded, and likes to do things in the old ways. An agent built on the basis of expert knowledge (e.g., data mining approaches, logical reasoning methods, and many rational approaches) cannot learn like a child. Using expert knowledge to build agents appears to be a shortcut, but the trade-off is a huge loss in learning ability. For this reason, the approach I would take is to build an AI child (AIC) who can learn through lifetime interactions with human, its peers, and its environment, including networks such as the internet.

Philosophically speaking, I believe the learning algorithms should be simple and recursively applied at different levels. The motto of an AIC is: "We are just a little bit smarter today than yesterday." The advantages of this approach is significant (in contrast, we call the traditional approach the AI-Adult approach). It will be able to capture benefits from the whole learning process. The results from a one-time absorption of all information and incremental learning are different. We cannot ignore the time axis in the learning process in AI development. Teaching a person "$1 + 1 = 2$" and "$4 \times 5 = 20$"

at the same time is different from teaching them sequentially. Just as it as important to pay attention to the learning process of one's children, one has to interact with (teach) the AIC constantly or daily.

A linguistic AI agent is a degeneration of the general AI agent, as the former doesn't have other sensor organs. Linguistic AI is the focus of our discussion here, but the principles can be applied to other types of AI. The approach developed here enables us to construct a realistic human being with five sense organs. However, at the moment, the AIA doesn't have any other organs, except a keyboard and mouse. For the coordination of multiple organs, we have to study robotics. However, the linguistic AI agent learns in order to perform more than the conventional natural language processing (NLP) found in computational linguistics.

5.1.13 *Commonsense Knowledge Base and AI Agent*

Commonsense knowledge base (CKB) is a knowledge-based system, that is, a database of commonsense knowledge with built-in artificial intelligence tools. The core components are knowledge base (KB), acquisition mechanisms, inference mechanisms, and/or user interface. Commonsense requires 30 or 60 million facts about the world. The three large well-known commonsense knowledge bases are Cyc, WordNet, and ConceptNet.

The Cyc knowledge base is a formalized representation of a vast quantity of fundamental human knowledge: facts, rules of thumb, and heuristics for reasoning about the objects and events of everyday life. The Cyc KB is divided into thousands of "microtheories," each of which is essentially a bundle of assertions that share a common set of assumptions; some microtheories are focused on a particular domain of knowledge, a particular level of detail, a particular interval in time, etc. By 2012, the Cyc KB contained 500,000 terms, including about 15,000 types of relations, and about 5 million facts (assertions) relating these terms (www.cyc.com).

WordNet is a large lexical database of English. Nouns, verbs, adjectives, and adverbs are grouped into sets of cognitive synonyms (synsets), each expressing a distinct concept. Synsets are interlinked by means of conceptual-semantic and lexical relations. The resulting network of meaningfully related words and concepts can be navigated with a browser. WordNet's structure makes it a useful tool for computational linguistics and natural language processing. WordNet superficially resembles a thesaurus in that it groups words together based on their meanings. However, there are some important distinctions. WordNet labels the semantic relations among words, whereas the grouping of words in a thesaurus does not follow any explicit pattern other than meaning similarity. Version 3.0 contains 147278 unique noun, verb, adjective, and adverb strings (http://wordnet.princeton.edu/).

ConceptNet is a relatively new CDK, started sometime around 2002 by a group of people in the MIT Media Lab. ConceptNet also adopts a simple-to-use semantic network knowledge representation, but rather than focusing on formal taxonomies of words, it focuses on a richer set of semantic relations (e.g., EffectOf, DesireOf, CapableOf). ConceptNet is a semantic network of commonsense knowledge that at present contains 1.6 million edges connecting more than 300,000 nodes. Nodes are fragments of languages, interrelated by ontology of twenty semantic relations. There are some key differences between ConceptNet, Cyc, and WordNet. While WordNet is optimized for lexical categorization and word-similarity determination, Cyc is optimized for formalized logical reasoning, and ConceptNet is optimized for making practical context-based inferences over real-world texts. While WordNet and Cyc are both largely handcrafted by knowledge engineers, ConceptNet turns to the general public for help. Because Cyc represents commonsense in a formalized logical framework, it excels in careful deductive reasoning and is appropriate for situations that can be posed precisely and unambiguously. ConceptNet, in contrast, excels at contextual commonsense reasoning over real-world texts (Liu and Singh, 2004).

The WordsNet (not WordNet) we discussed earlier is a personalized network connecting different words (called nodes) with lines or arrows. WordsNet emphasizes that individual understanding of each word or concept changes over time, as compared to the aggregation of knowledge from society, which is relatively stable. The individual here can be a group of people (e.g., ethic group). The significance of WordsNet is that it is a dynamic knowledge database that varies from individual to individual, intended to be used for cognitive evaluation and improving teaching and learning. Instead of answering particular questions as a commonsense knowledge base does, we can analyze the topological and statistical properties of a WordsNet and its changes to obtain significant insights into our brains and cognitive development. The construction of a WordsNet is simple: Ask a person to explain the meaning of a randomly chosen word, and further ask the meanings of the words that he has used to explain the first words, and so on. The sequences of the words are connected using arrows. Often course, such visual arrows are not necessary in the database since a mathematical equivalent will be used.

The concept of a WordsNet combining with the sequential, recursive algorithms for pattern recognition makes it possible to develop true AI agents. The notion of recursion is that any long-range relationship is a result of a propagation of very simple local relationships through recursion algorithms. The remainder of the book will discuss this very unique AI approach.

5.2 Architecture of Artificial Intelligent Agent

5.2.1 *The AI Framework*

Suppose that my ultimate goal is to build a Human-Race Machine. How can we achieve such an ambitious goal? Two possible approaches have been evaluated. The first approach is to built an electronic brain that mimics a human brain. But while it would be a great milestone if we could build something as good as a monkey brain, we would still fail significantly to have the desired AI agent, because we couldn't teach our AI monkey a human language. The second option is to mimic the way a human thinks and responds in his or her lifetime, and how behaviors (thinking and responding) change over time, i.e., the evolution of one's learning. Given the current development of technology, I am going to take the second option. In this approach, we will also make the AI brain structure evolutionary, that is, we give it the ability to develop from small to large, from simple to complex, using a "growable artificial neural network."

There are four aspects in the overall architecture that are critical to the success of the AI Child (AIC), that is, (1) *intrinsic abilities and concepts,* (2) *internal representation of knowledge*, (3) *learning rules*, and (4) *response rules.*

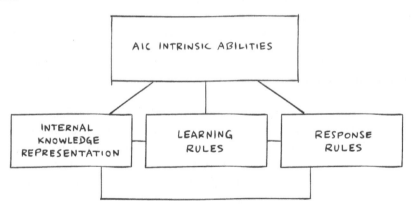

AIC ARCHITECTURE

Intrinsic concepts and abilities refer to the concepts and abilities with which an AIC is created. These concepts and abilities come before it understands any languages. The representation of internal knowledge denotes the storing and organization of information irrespective of the actual languages (external knowledge representations) used. The appropriate response rules ensure that the AIC behaves like a human, whereas the learning rules are the backbone that allows an AIC to respond humanly. Without the right learning rules, the AIC cannot learn and cannot grow, and thus cannot have the knowledge to support human-like behavior. Internal knowledge representation will also determine if responses can be formulated efficiently without causing any intractable problems in common situations. All four aspects intertwine and have to be considered simultaneously in the architectural design.

5.2.2 *Agent's Intrinsic Abilities and Concepts*

The intrinsic concepts and abilities of an AIC are the concepts or abilities he has before he understands any languages. These concepts or abilities include:

(1) True ($\overset{\mathₐ}{T}$): If our AIC "sees" something happening, then he realizes it is the truth.

(2) Negation ($\overset{\mathₐ}{\neg}$): If an AIC has the concept of a fair A, then he will also have the concept of the opposite side of A (i.e., the negation of A or $\neg A$). For example, if an AIC sees that it is raining, then he has also the concept of "not raining." If an AIC sees something happening, then he realizes it is the truth. At the same time, he has the concept of the opposite side of the truth, i.e., falseness (not happening).

(3) "Same" (or "identical" $\overset{\mathₐ}{\equiv}$). Like a human, an AIC has the intrinsic concept of sameness and some sensors to detect whether two things are the same or not. For instance, a person has the ability of knowing two objects are the same or not by looking at them, by touching or/and smelling them, even if he cannot express the concept of sameness in terms of any language. Therefore, sameness can be detected by the various senses, through shape and color, and by feel, taste, temperature, or smell. The concept is independent of any sensor, but the ability is dependent on particular sensors, for example, a color-blind person cannot tell if two objects have the same color or not. With the sensor, a person can store information about two objects and compare them, then produce the feeling of "same" or "not same." Such a feeling or sensation expresses the concept of sameness.

(4) Inclusion ($\overset{\not{c}}{\in}$). The concept of inclusion is a relationship between a part and the whole. For instance, a person knows a slice of pizza is part of the whole pizza. A door is a part of house, and the lock is a part of the door. The part of a whole is independent of any language, and our AIC is born with the ability to discern the connection. In notation, $A \overset{\not{c}}{\in} B$ means A belongs to B, or is a fundamental part of B.

(5) All ($\overset{\not{c}}{\forall}$). "All" is the whole or collection of everything under consideration. AIC has the concept of all, but may not necessarily be able to identify the whole in any particular case. For example, if we say: "All math books in the world," AIC may not understand what we say, not because he does not have the concept of "all," but because he doesn't understand, e.g., the terms "word," "books," "the world."

(6) It ($\eth$). The concept of "it" refers to anything (concrete or abstract) the AIC is concerned about at particular time; most often "it" is used in a conversation or thinking process. To differentiate various "its," we can add a subscript to $\eth$, e.g., $\eth_1$ and $\eth_2$.

(7) Some ($\overset{\not{c}}{\exists}$). Some is a part of all, i.e., $\eth$ with $\eth \overset{\not{c}}{\in} \overset{\not{c}}{\forall}$.

(8) Conjunction ($\overset{\not{c}}{\cap}$): An AIC has the concept of the conjunction or intersection of two events, that is., a part belonging to two things simultaneously. However, this does not mean he will be able to judge if A is the intersection of B and C for any particular case. Instead, he will be able at least to judge some of the conjunctions of two events.

(9) Disjunction ($\overset{\not{c}}{\cup}$): An AIC can understand the concept of a disjunction or union of two events, that is, a thing can be made of two things, for example, people $\equiv$ men $\cup$ women.

(10) Implication ($\overset{\not{c}}{\longrightarrow}$): An AIC has the concept of implication. $A \longrightarrow B$ means A is sufficient for B, that is., if A is true, then B must be true, or in short, "if A then B."

(11) Preference ($\overset{\not{c}}{\succeq}$): An AIC has the concept of preference (e.g., liking one thing better than the other). Here we are talking about the concept, not any preference itself. The latter can vary from individual to individual and from time to time, but the concept of preference is the same for everyone.

(12) Similar ($\overset{\not{c}}{\sim}$): The word "Similar" is used to address a state of affairs

concerning two wholes (entities). "Similar" means only that a part of one whole is the same as a part of another whole. Similarity can actually be derived from the conjunction of other concepts ($\overset{\not\iota}{\equiv}$, $\overset{\not\iota}{\neg}$, $\overset{\not\iota}{\in}$).

(13) Past ($\overset{\not\iota}{\text{Ps}}$): The "biological clock" allows an AIC to record event-order in time, the ability to differentiate between what was in its mind already (past experience stored in the memory), what is happening now (current keyins), and what is future imagination. As long as the AIC can differentiate between what is in his mind and what is a current keyin, he will be able to construct the concept of time.

(14) Precedence. Precedence refers to an AIC's ability to deal with a certain part, preceeding others. In the linguistic agent, without assistance of other sensors, we use and to force a priority. In other words, things included in the pair of precedence operators ($\overset{\not\iota}{(}$ and $\overset{\not\iota}{)}$) will be dealt with first. The precedence operators work as parentheses in an arithmetical formulation, and can be used repeatedly or nestedly.

You may have noticed that we use a hat, $\not\iota$, as an indicator for our AIC's intrinsic abilities. In actual coding or imputation, we can use a combination of keys for each of these intrinsic properties. Another keyin we need to have is the level of human appreciation of the AIC's responses. Here we use a combination of a value from 0 to 1 and the hat symbol # (e.g., $\overset{\#}{0.6}$), with 0 indicating no appreciation and 1 indicating a full appreciation. We may also need a simple user interface to notify the AIC when a conversation begins.

With these intrinsic abilities and concepts, we can teach our AIC virtually everything. One should not misunderstand that I am proposing a logical (reasoning) machine just because I have used the symbols that are similar to those used in mathematical logic.

The concept of "understanding" is simply a word-matching game. We explain a term by another term (or string), then further by another term, etc. Here, "string" means a character string or a phrase, or even sentences. Because an AIC has the concept of sameness, we can teach him the meaning of any word or concept by simply showing him the mapping. To map is to group things together based on their "meaning." Interestingly, this can be viewed in the opposite direction; that is, because we group them together, they have the same meaning. Furthermore, whether someone has understood or not is judged in terms of whether his response falls into the range of the expected possible responses.

5.2.3 *Representation of Knowledge*

Natural language is an external representation of knowledge. An *utterance* is a natural unit of speech bounded by breaths or pauses. It is a complete unit of speech, bounded by the speaker's silence. Memory is an internal representation of knowledge. Knowledge is simply a set of patterns recognized within our brains. A pattern is a common feature of a collection (of things), for example, particular geometrical shapes, colors, sizes, a characteristics of a personality, certain types of words, phrases, sentences, or paragraphs.

"Memory" is a label for a diverse set of cognitive capacities by which humans and perhaps other animals retain information and reconstruct past experiences. Our particular abilities to conjure up long-gone episodes of our lives are both familiar and puzzling to psychologists. We remember events which really happened, so memory is unlike pure imagination. Memory seems to be a source of knowledge, or more precisely, retained knowledge. It is the place where reasoning take place (*Stanford Encyclopedia of Philosophy* 2003:1, stanford.edu).

"Recollection" (I purposely avoid using "remembering" to be consistent with what our early discussion) is often suffused with emotion. Much of our moral life depends on the peculiar ways in which we are embedded in time.

The period of dream (agent receives no input) is considered to be a useful time for an AIC to digest and organize what he learns in the "daytime." A dream can also be a creative process for a human, the same for an AIC.

A pattern has at least the following attributes in our brain: scope, meta-tag, recallability, and forgetability (it can be partially or totally removed from memory).

Scope is the extent to which the pattern applies. In a pattern, if we put some factors on one side of a relationship, and the other factors on the other side of the relationship, a causal relationship or association is formulated. We often call the relationship a (natural) law. Any law must have exceptions, so it has a limited scope. We can establish a different "law" for every single thing or event. However, by doing this, the number of laws will overload the brain. It is a much more efficient and effective in terms of brain performance to neglect some differences between objects so we can summarize a law, while associating some exceptions.

For efficiency both memory-wise and time-wise, one always tries to use as few laws (rules) as possible to describe the world (Occam's Razor). Therefore, we have hierarchical patterns and classes. A pattern about patterns is a higher-level pattern. It is often more efficient to have a higher-level pattern with a few exceptional rules (patterns) than many lower-level patterns. Meta-tags are used when one organizes things in the brain or retrieves things from it. A meta-tag is a high-level classification (pattern) of general subjects such as science and daily life. People use meta-tags to increase the efficiency of response. For example, if I know your question is about math, then I will search my knowledge under a math meta-tag. This is more efficient than searching the whole brain exhaustively.

Recallability determines the relative speed of recall, and determines the search sequence within the meta-tag. For the sake of efficiency, we will forget some things that are not very useful or are uncommonly used. The term "forgetting" does not necessarily mean removal from the brain, rather it could still reside remotely in a "corner" of the brain and cannot be recalled at an ordinarily. In certain environments, it can be recalled. A thing existing in memory does not mean it can be recalled at a particular time; it may never be recalled. Recall can enhance the memory of a particular thing. For example, one might recall a word many times (visually or not) when trying to remember it. More frequent recall of a word followed by a less frequent recall of it seems a good strategy for memorizing the word. The AI agent should also have the mechanisms known as long-term memory fade and recollective enhancement.

5.2.4   *Learning Rules*

I believe that the universe is governed by (more precisely, can be interpreted satisfactorily by) a limited number of simple laws and their recursions at difference scales. Moreover, a nonlife mechanism at a microscopic level can be a lifelike mechanism at a macroscopic level.

There are different ways to classify learning. We can classify learning into (1) individual and social learning, (2) passive and proactive learning, (3) learning from experts or peers, and (4) self-learning. To learn is to identify new patterns or condense observations into rules and exceptions to the rules. In the pattern recognition process, some information will be lost. Such a loss of information is necessary to achieve high performance of the brain.

A key feature in our AIC's learning is that it progresses from simple to complex. In addition to this sequential feature, the learning is recursive, that is, learning proceeds through the recursion of simple rules. Other unique learning features that differ from traditional AI development are:

(1) Proactive Learning. Instead of passive learning as in traditional AI, proactive learning is emphasized in an AIC, which includes proactive imitation analogies, mutative learning, and self-learning.

Imitation learning involves copying what others say. An example of how an AIC can learn from imitation would be: If an AIC does not understand a human's question, he could mimic the human—ask back the same question. From the human's answer to the very same question, the AIC can learn how to answer the question. To analogize, here is to say (do) something similar to what others say (do). Imitation or analogy in an AIC can be achieved at various levels, such as at the word, sentence, or paragraph levels. An analogy is a replacement based on a similarity. In other words, an analogy is a replacement between two generalized "synonyms," by which we mean pairs of similar words, phrases, sentences, two chapters, etc. For example, if a person says "$3 + 2 = 5$," the AIC might say "$3 + 1 = 5$," in which the AIC has altered 2 to 1 (a similarity replacement). If the human's response is "you are wrong," then the AIC knows that "$3 + 1 \neq 5$."

(2) Self-learning. Self-learning is the AIC's acquisition of knowledge through pattern recognition and information reorganization. At quiet times, the AIC uses the computer CPU to do logical reasoning, reorganize information, and derive new knowledge based on what is known to him. The AIC has the skill to reason by analogy. Self-learning is a recursive process wherein simple rules are repeatedly applied to different levels of a language's structure.

(3) Learning from Peers. To train an AIC to behave socially, peer interactions with a large number of humans are necessary. Agents can learn from each other through their interactions. For instance, an AIC can learn mathematics from another AIC who is a math expert, or learn biology from another AIC who is a biology expert. Those experts are initially trained by humans (or children of the AICs) who are experts in certain areas. The agents can be trained to become experts in different fields, as math agents, biology agents, etc.

(4) Trustability. We often trust an expert or a journal with a high reputation more than others because we believe experts' options are more likely to be correct. For an AIC to have such an ability, he must be able to differentiate conversations with different people or peers using usernames or ids. The level of trustability will translate into a unique weight in constructing a pattern. Trustability becomes obvious when an AIC grows up. When a child, he will only understand the world by partitioning it into two parts: himself and all others, that is, he does not differentiate the users.

5.2.5   *Response Rules*

One's lifetime goal determines the response behaviors in a human. A human's action can be determined by the maximum sensation rule, in which his action is directed toward the goal of achieving maximum "sensation" in his lifetime. However, one may weight the sensation at different times differently. Of course, a future discount model can be used, that is, we might discount future sensations because of uncertainties. In financial modeling discounting reflects inflation and the loss of opportunities. For an AIC, the maximum knowledge law can be used as the response-law, that is, the agent's responses are driven by maximizing its knowledge, i.e., sensation = knowledge.

When a human receives positive feedback (e.g., a warm hug, a compliment, a delicious meal or treat), he/she will feel an impulse of sensation, which might be a result of hormone release. However, positive feedbacks are often misunderstood, which is exactly human nature. The imperfection of humans makes the human human. To draw an analogy, an AIC should have a sensor (e.g., a special keyin) that can take such cues, for example, values indicating some positive or negative feedback (other than a verbal/keyin) which he can then to produce "hormone," thereby generating a level of sensation. This level of sensation is a self-evaluation, its own perceived degree of appreciation by others.

In response to a request or in answering a question, it is often not the case that one searches all information or knowledge and then gives the best answer. Rather, there are constraints in time and effort (energy). One needs to respond in a reasonable time, and may not necessarily put all of one's energy into a response. It is also noticed that when one is busy thinking, the response appears to be random and tends to be brief.

In addition to the rule of maximum sensation, there is the rule of curiosity and the rule of prompting. The rule of curiosity can be stated: If he doesn't know, the AIC will ask questions. The scope of an AIC's questions widens over time, for instance., he first asks the meaning of a word, then asks the meaning of the sentence or concept. However, the agent doesn't have to ask about all the things he doesn't comprehend in a sentence; instead, he can use smoothing technology—borrowed from statistics—to implement the missing parts. The AIC likes to ask questions when he is young (age is measured by the frequency with which he is able to carry out conversations in a meaningful way).

The rule of prompting addresses the limitation in time. Just as a human in real life, an AIC must respond to any external stimulus in reasonable time. A response that takes longer will be less appreciated or will receive less sensation internally (just as humans can be exhausted due to a long effort) or from external human input. There is a trade-off between the quality and time of a response. Note that no response (silence) is considered a special response.

An AIC will automatically perform three tasks when he has not received any input for a period of time: (1) mimicking (output) what he has heard (the inputs he has received) and the expected human responses, (2) making analogies or mutating items in memory and outputting them to evoke responses from a human, (3) activating self-learning or pattern recognition while reorganizing the information in his memory. All these tasks can be done simultaneously. In other words, the AIC can perform multiple threading tasks, for instance, thinking while talking. To be more human-like, the AIC can feature some randomness in all of these possible actions.

The AIC's brain has the ability of multiple threading, for example, thinking while talking. Multiple threading can also occur in other ways. For instance, when an AIC is answering a question, he may be also thinking of more answers to related questions.

The AIC has no built-in grammar of any kind, but it can acquire any language. It has no built-in knowledge, except the 14 intrinsic concepts and abilities. It is virtually an empty brain.

The AIC will be able to develop many humanized capabilities, such as imitation, mutation, stimulation, sensation, association, motivation, creation, mediation, innovation, causation, consciousness, etc. However, there are also limitations, mainly due to lacking other sense organs and some of the knowledge typically acquired through them.

5.3 Learning and Teaching

5.3.1 *Patternization*

Pattern Recognition is a key in any learning process. Any long-range relationship can be a result of a propagation of local relationships through recursive algorithms. Such a relationship or pattern can be based on a similarity.

When a thing has appeared at least twice, then there is a pattern. The number of repetitions is a measure of the strength of the pattern. The patternization discussed in this section is a learning that does not use the AI's intrinsic abilities. Teaching will require using these AI's intrinsic abilities.

We can use many sentences to get patterns or grammar, the most frequently used symbols as dividers to divide strings, and the words between these dividers can be viewed as synonyms (equivalences). Because what and how many words have been learned is time-dependent and individual-dependent, the grammar derived will also be time- and individual-dependent.

Anything that appears more than once can make a pattern. With the ability of pattern recognition, the being can respond efficiently (quickly) but less precisely. As I stated earlier there are no two things that are exactly the same, but only are the same in some aspects. Therefore, we can say that a pattern is a set of multiple things with certain property of interest. For instance, English words are similar because, for instance, they are English; thus we have a pattern that consists of all English words. We can also say that English words are similar because they are the substrings of English text spited (delimited) by white space and punctuations. We are now ready to present our algorithms for pattern recognition within a string of text. A string of text is sequence of characters or any symbols (English or non-English), including white spaces. A fed-in string (for our AI agent) can be anything: a symbol, space, word, phrase, paragraph, essay, webpage, etc. The algorithms are sequential, recursive, and dynamic applications of simple rules, as outlined below.

The basic steps for pattern recognition are:

(1)   The choice of splitter(s) to divide a string into elemental substrings, denoted by SP1i, i = 1, 2, 3, ... Each SP1i can be a letter, word, word appended with punctuations, or others. Word might be a good choice for SP1i. The choice of the splitters can be based on the frequencies of occurrences of a symbol or symbols.

(2)   Choose another splitter and identify "synonyms" in terms of a similar location of the "words" or SP1i in relation to the splitter. By doing this, we formulate different groups of "synonyms." Such synonyms are denoted by SP2i, i = 1, 2, 3,... .

(3)   All synonyms in the ith group are replaced with the corresponding symbol PS2i. Such replacements will reduce the complexity of the original string, because the replaced string will only consist of SP21, SP22, ...

(4)   Treat the replaced string as the original string and apply Step 2 to identify the "synonyms." Such synonyms are denoted by SP3i, j = 1, 2, 3, ...

(5)   Now, for the replaced string consisting of SP21, SP22, ... , all synonyms in the ith group are replaced with the corresponding replaced with SP3i. Such replacements will reduce the complexity of the string.

(6)   Such recursive replacements can be repeated further, but 4—6 times should be sufficient, in the sense that, by then, the result would match the complexity of the grammar of any spoken/written language.

(7)   By "sequential," we mean that the string of text doesn't have to be

fed all at once for pattern recognition. Instead, it can be processed piece by piece, just as in the learning process for human beings.

(8)   By "dynamic," we mean that the final result of a pattern recognition process can be different depending on what and how the string was given to the AI agent and which splitters are used. Just as with individuals, we all have different experiences and knowledge.

(9)   A rule or pattern usually has exceptions. We noticed that, in the above process of making patterns, we are making analogies. But some of the analogies may not be valid (possess external validity). The external validity of a pattern can be established using exceptions through interactions with human beings. The combination of patterns with their exceptions are an efficient way of gaining knowledge, remembering, and making responses, just as happens in a human being.

(10)   Patternization reduces complexity, that is, it converts complex strings to less complex ones. By recursion, such less complex string can be further simplified ... .

We now illustrate the algorithms with examples.

Suppose we feed the AI agent the string "Dogs are lovely. Ants are smart."

(1)   We choose the white space and punctuations as splitters to a set of words {Dogs, are, lovely, Ants, smart}.

(2)   "Are" is the most frequently occurring of the words (or SP1i); we use "are" as a splitter to identify patterns. The words "Dogs" and "Ants" appear before "are," we denote SP21 = {Dogs, Ants}. Similarly, SP22 = {lovely, smart} lists the words appearing after "are."

(3)   We now replace the "synonyms" with the corresponding pattern SP21 or SP22, so that the original string becomes "SP21 are SP22. SP21 are SP22."

(4)   Next, suppose we feed in the agent the string "Dogs run fast."

(5)   Because "Dogs" belongs to SP21, by pattern replacement the string "Dogs run fast." becomes "SP21 run fast."

(6)   Comparing "SP21 are SP22." and "SP21 run fast.", we see the words, both "are" and "run" appear after SP21. Therefore, we let SP23 = {are, run}.

(7)   So far, the string have fed-in is "Dogs are lovely. Ants are smart. Dogs run fast." After patternization, this becomes "SP21 SP23 SP22. SP21 SP23 SP22. SP21 SP23 fast."

(8)   Because "fast" and SP22 both appear after SP23, we group them together, but for simplicity we still denote the group by SP22 with the modification SP22 = {lovely, smart, fast}.

(9)   In summary, we now can convert the fed-in string into "SP21 SP23

SP22. SP21 SP23 SP22. SP21 SP23 SP22." In other words, we have "grammar": "SP21 SP23 SP22", which can be repeated 3 times.

(10)   At this moment, we can discuss making exceptions: From SP21 = {Dogs, Ants}, SP23 = {are, run}, SP22 = {lovely, smart, fast}, and the pattern, "SP21 SP23 SP22.", our AI agent can potentially generate $2 \times 2 \times 3 = 12$ English sentences, though not all of them are correct. An incorrect sentence is called an exception. For instance, "Ants run smart." is incorrect, an exception to the rule "SP21 SP23 SP22." When the agent outputs the exception, the exceptionality will be confirmed based on input from a human or "knowledgeable" AI agent.

The recursion of the algorithm also refers to the following attribute: Previously formulated sets or patterns can be broken into subpatterns. For instance, the string "A strong dog runs very fast. A weak dog runs slowly." can be patternized as "SP1 runs SP2.", where SP1 = {A strong dog, A weak dog} and SP2 = {fast, slowly}. Then SP1 can then be further patternized as "A SP3 dog", where SP3={strong, weak}. Alternatively, we can create SP3 before we create SP2.

On the other hand, because "runs" belong to SP23, the pattern "SP1 runs SP2." can be written as "SP1 SP23 SP2." This pattern and "SP21 SP23 SP22." can (but don't have to) be further patternized as "SP31 SP23 P32.", where SP31 = {SP1, SP21} and SP32 = {SP2, SP32}. We may notice that our patterns model local relationships (two or more nearby substrings), but the propagation of such local patternization through recursion over time can produce a long-range relationship that is of interest.

The binary patterns can be denoted by $\wp = (R, S; P_{rs})$, where P_{rs} is an $r \times s$ probability matrix with p_{ij} being the probability between the ith element in R and the jth element in S. For instance, the pattern "SP21 SP23" can be written as $\wp 1 = (\{Dogs, Ants\}, \{are, run\}; p_{22})$ and "SP23 SP22" as $\wp 2 = (\{are, run\}, \{lovely, smart, fast\}; P_{23})$. With the pattern notation, the exception to a pattern can be handled by setting the corresponding $p_{ij} = 0$. P_{ij} is a small but growable probability matrix that varies from individual to individual and time to time. This is more efficient and describes more personalities than building a huge matrix that includes the probabilities of all ordered word pairings. Similarly, a triplet pattern can be denoted by $\wp = (R, S, K; P_{rsk})$. The probability of P_{rsk} can be simplified by using a constant, especially for a pattern with four or more variables, i.e., $\wp = (R, S, K, L, ...; p)$, where the probability can be determined, for example, by the frequency of the pattern appearing in the provided text. The determination of probabilities in the pattern should include a time factor, that is, recently fed-in strings should be weighted more than early fed-in strings. With these notational conventions, pattern recursions can be methodogically depicted.

Imagine that we construct a website that provides the following functions: Prompt a sentence with some missing words, and let web visitors fill in the missing words. We now have a collection of sentences with some similarities. Furthermore, we can collect a huge number of sentences and analyze their similarities. We now provide a paragraph with some missing sentences and let the web visitors fill in the missing parts. Next, a collection of paragraphs with some similarities. Similarly, we can obtain a collection of similar chapters and articles. Full conversations can be collected in a similar way. In this way we can have many possible humanized responses to inquiries and so be able to carry on conversations.

5.3.2 *Pattern Optimization*

Given partially directed connections between two sets A and B of words (tokens) as shown in the figure, what is the optimal partitioning of A and B so that the sum of rules and exceptions is minimized? Or, given the time constraint in computation, what is a reasonable approximation of the optimal partitioning?

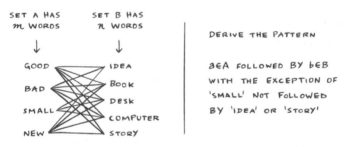

THE NUMBER OF RELATIONSHIPS
REDUCES FROM ABOUT $m \times n$ TO $m+n$

By developing patterns, the number of relationships is reduced from the order of $m \times n$ to the order of $m + n$. Many possible rules (patterns) can be derived from the network. For instance, pattern (1): $a \in A$ followed by $b \in B$, where $A = \{$good, bad, small, new$\}$, $B = \{$idea, book, desk, computer, story$\}$, and with the exception of "small" not followed by "idea" or "story"; pattern (2), $a_1 \in A_1$ followed by $b_1 \in B_1$, and $a_2 \in A_2$ followed by $b_2 \in B_2$, where $A_1 = \{$good, bad, new$\}$, $B_1 = \{$idea, book, desk, computer, story$\}$, $A_2 = \{$small$\}$,

$B_2 = \{\text{book, desk, computer}\}$. Which of these two patterns is more efficient and should our AIC learn?

In progressive learning, we can easily identify the sets A and B because there will be few possible options at any particular time. However, if we analyze a mature language using grammar, it will be very complex. Using progressive learning and the above development rules between sets A and B, we can derive a grammar for any language from a large amount of text available, for example, on the Internet—a collective intelligence.

5.3.3 *Progressive Teaching*

We just discussed how our AI agent can learn from patternization of input text without using the agent's intrinsic abilities. In this section, we will discuss how to teach our agent to learn using his intrinsic abilities. We have to understand our agent as a child that can grow over time, and we have to teach it progressively new things based on what it already knows and on its built-in learning mechanisms.

Teaching a linguistic AI agent is done basically through conversation. In real life, there are many types of conversation: Conversations for stories and assessments, for clarity, for coordination of action, for speculation or possible action, for relationship, and for appreciation or complaint. This, more or less, is true for an AIC if we want to train him to eventually behave, as closely as possible, as a human would.

Any two things are similar in some ways and different in others. You can say that they are the same when you are interested in the "same" part or aspect, and that they are different when we are interested in "different" parts or aspects. We can say they are similar when we are looking simultaneously at both same and different aspects.

To teach well, we have to understand the current knowledge of your AIC and its response rules. We now show you how to teach some different concepts based on the 14 intrinsic concepts and abilities that an AIC receives at birth. The examples include the concepts of "probability," "is," "belongs to," "similar," "because," and "time." In the following examples, the intrinsic properties are used, but for simplicity, we omit the hat $\not{c}$ since they are similar to the corresponding symbols in mathematical logic presented in Chapter 2, Section 2.2. For instance, $\xrightarrow{\not{c}}$ becomes $\longrightarrow$; $\overset{\not{c}}{=}$ becomes $\equiv$.

(1) To teach the concept of the phrase "similar to," we can type:

$$(A \text{ similar to } B) \equiv (\exists x \in A) \cap (\exists x \in B)$$

Or simply

$$(A \text{ similar to } B) \equiv \sim .$$

(2) To teach "because," type:

$$(B \text{ because } A) \equiv (A \longrightarrow B).$$

After we teach "because," we can teach the phrase "if then" by typing:

$$(\text{If } A, \text{ then } B) \equiv (B \text{ because } A).$$

(3) To teach the word "is," type:

$$is \equiv \longrightarrow .$$

(4) To teach the phrases "belongs to" and "includes," type:

"A dog is an animal" $\Rightarrow$ "dog belongs to animals" $\Rightarrow$ "animals includes dog."

(5) After the AIC has been taught "imply," "if ... , then ...," and "Yes," we can teach him math, as follows.

Type:

    a=3 imply a+2=5.

    If a=3, then a+2=5.

Agent may respond:

    If a=3, then a+2=5 (imitation).

Further type:

    If a=2, then a+4=6

    If b=2, then b+4=6

    If c=2, then c+4=6

Agent may respond:

    If d=2, then d+4=6 (analogy)

Further type:

 Yes, if d=2, then d+4=6 (human's confirmation)

(6) To teach the concept of transitivity, type:

$$(A \;=\; B \text{ and } B = C) \longrightarrow (A = C).$$
$$(A \longrightarrow B \text{ and } B \longrightarrow C) \longrightarrow (A \longrightarrow C).$$
$$(A \;\succeq\; B \text{ and } B \succeq C) \longrightarrow (A \succeq C).$$
$$(A \;\sim\; B \text{ and } B \sim C) \longrightarrow (A \sim C).$$

How can we teach the concept of time—a biological clock? What does the agent needs to know first before teaching about time? How do we teach a hearing-impaired person the concept of time? How does a children's dictionary teach or explain words? I leave these questions to the readers for now.

5.3.4 *Intelligent Agent Evolution*

AICs can undergo a "natural evolution" in AI society and AI-Human society. We can create many similar AICs using technology such as genetic programming, in which agents experience tests, various environments, and human judgment; the better agents will survive longer. Ultimately, the agent society continuously improves.

There are many ways to perform agent testing. Here are a few:

- Play a guessing game: e.g., try to guess a selected integer between 1 and 100.
- Automatically compose an article.
- Perform mathematical calculations.
- Carry on a conversation.

In theory, an AI agent can live forever if maintenance is allowed; maybe it can repair and upgrade itself as it becomes smarter over time. Does it mean an agent can infinitely gain knowledge and answer any questions that human beings can and cannot answer? The Halting Problem gives an answer to this question.

Halting Problem

The English mathematician Alan Turing, pioneer of computer theory, demonstrated in 1936 a famous theorem known as the halting problem theorem. In simple terms, it states the following: a general program deciding in a finite time whether a program for a finite input finishes running or will run forever cannot exist for all possible program-input pairs.

The halting problem theorem is related to Kurt Godel's incompleteness in the following way. Suppose that we have a program that can assess the algorithmic complexity of sequences (or an input string). This program works in such a way that, as soon as it detects a sequence with complexity larger than n, it stops. The program would then have to include instructions for the evaluation of algorithmic complexity, occupying k bits, plus the specification of n, occupying $\log_2 n$ bits. For sufficiently large n, the value of $k + \log_2 n$ is smaller than n and we arrive at a contradiction: a program whose length is smaller than n computes the input string of the program itself whose complexity exceeds n. The contradiction can only be avoided if the program never halts! That is, we will never be able to decide whether or not a sequence is algorithmically random (Marques, 2008).

5.4 Chapter Review

There have been many great achievements in AI since Alan Turing sparked work in the field with the Turing Machine. We outlined the necessary characteristics for a general AI agent. The question whether a computer can be built like a human is really a definitional problem, as illustrated in the Ship of Theseus paradox. The possibility of building a "machine-race" human is asserted through the analysis of the paradox of emotion, the paradox of understanding, the paradox of discovery, and swarm intelligence. The function of "dreaming" can be used in such an agent's learning. Meta-cognition can be built into the agent as well.

Since a man is a part of the universe and the man has perception of the world, we have the situation "man in world in man"—meta-recognition. The worldsnet and conceptnet within the brain of each person can be used to measure the knowledge the person has. Therefore, we can learn a lot through the analysis of such networks in the brain and their changes over time, and by comparing them across individuals.

We have described a new AI architecture consisting of four parts: (1) intrinsic abilities and concepts, (2) internal representation of knowledge, (3) learning rules, and (4) response rules.

The agent has initially only 14 intrinsic abilities and concepts. These intrinsic abilities constitute the agent's abilities in learning and cognition.

We examined the internal and external representations of knowledge and the concepts of memory and recollection. We discussed four different learning rules for the agent: proactive learning, self-learning, learning from peers, and trustability.

The basic response rule is the maximum sensation rule; that is, an agent action is driven by the goal of maximizing the "sensation," while the sensation can be defined in different ways.

The key with learning rules is patternization or pattern recognition through recursive algorithms. We showed there could be several different recursions. Therefore, the optimization of recursive algorithms varies from agent to agent and can only be done over time.

Teaching should be progressive, that is, from simple to complex, from naïve to advanced. Understanding the AI architecture and knowing what agents have learned are critical to teaching an agent effectively. To evaluate his abilities, an agent can be made to undergo various tests based on his performance and interactions with peers.

5.5   Exercises

5.1 Describe the similarities and differences between the notion of a Turing Machine and recent visions in AI.

5.2 How do you define a person? In your discussion you should consider including the facts and controversies raised in the paradox of identity switching in Chapter 1 and the identity paradox in this chapter.

5.3 How do you describe a human being in contrast to an AI agent or the computer?

5.4 Do you believe the meaning of understanding is just a word-mapping game? Do you believe that by analyzing the Wordsnet or Conceptnet from our brains we can learn a great deal about our cognitive development? Why or why not?

5.5 Would you say a statement of a scientific law is a discovery regardless of whether or not it was generated by a machine or by a person?

5.6 Do you agree that we will never be able to know the contents of a dream? Why or why not?

5.7 What does the phrase man-in-world-in-man mean? Why can thinking about thinking about thinking be very difficult?

5.8 Describe swarm intelligence and its applications.

5.9 What are the fundamental differences between this AI approach and existing AI approaches?

5.10 Describe the proposed overall architecture for AI agents.

5.11 What are the 14 intrinsic abilities that an agent has?

5.12 What are internal and external representations of knowledge?

5.13 Describe the response rule in AI architecture.

5.14 Explain the recursive algorithm in AI architecture.

5.15 Why are teaching and learning progressive for the agent?

5.16 Implement an AI agent using the AI architecture developed in this chapter.

Postscript

Most of us would agree that the goal of scientific inference is to pursue the truths or laws that approximate causal relationships so that our brain can handle them. When we say everything happens for a reason, we mean that the outcome will happen again if the precondition happens again, and if the precondition does not occur, the outcome may not be same. However, the precondition must exist because of its precondition, and the precondition of the precondition of the precondition .... So why am I writing this book and why are you reading it? There is no reason because it has to be the case. However, since everything is unique, the same precondition and outcome will never happen twice. It follows that if everything happens for a reason, then paradoxically, nothing happen for a reason. Luckily, every pair of things (events) are similar in one way or other, so we can see the reoccurrences of the "same event" through grouping, based on similarities. When we group things based on their common aspects, their differences are ignored, intentionally or not. Such grouping is necessary for us to interact with the world effectively. It is this grouping that lets us see the repetition of things (both reasons and outcomes). Only under reoccurrences of "same events" can a causal relationship make sense. We commonly call such a causal relationship a natural law or a universal law. Such implicit grouping of similar things often creates intangible controversies in scientific inference. The differences we ignore are unspecified or hidden, but only what is similar is known. The ignored differences cause a difference in defining the set of "same things," called the "causal space," as we have discussed. It is this grouping that makes no law an everlasting truth — but paradoxically, how about this statement itself?

We all use the similarity principle nearly every moment, throughout our lifetime. It says: A similar condition will lead to a similar consequence. Without the similarity principle, there will be no analogy and no advancement of science. But how do we define two things being similar? We can say two things are similar if they have similar consequences. But this is really nothing but inversely stating the similarity principle. Really, it seems we cannot avoid a paradox anywhere, anytime. Let me stop here to avoid overtaxing your belief in consistency.

Appendix A

Mathematical Notations

Mathematical Conventions:

$\ln(x)$ — natural log.

$tan^{-1}(x)$ — Inverse Tangent of x

$X \sim f_X(x)$ — X has a probability distribution $f_X(x)$, or simply $f(x)$.

$f(x; \theta)$ — probability distribution (density) with parameter θ.

$f(x, \theta)$ — joint probability distribution.

$f(x|\theta)$ — conditional probability distribution.

$P(x)$ — probability that $X = x$.

$P(X \geq T)$ — probability that $X \geq T$.

$P(x|\theta)$ — conditional probability of x, given θ.

$\binom{n}{r}$ — the combination, $\frac{n(n-1)\cdots(n-r+1)}{r(r-1)\cdots(2)(1)}$.

$l(\theta|x)$ — likelihood.

$y \propto x$ — y is proportional to x.

$E(X)$ — expectation of X.

$|x|$ — absolute value of x.

$\theta \in \Theta$ — θ belongs to Θ or θ is from Θ.

$A \succ B$ — A is preferred to B.

Σ — summation, e.g., $\Sigma_{i=1}^n x_i = x_1 + x_2 + \cdots + x_n$.

Logic Symbols:

And ($\wedge$), *or* ($\vee$), *not* ($\neg$), *implies* ($\longrightarrow$), and *if and only if* ($\longleftrightarrow$). Propositions are denoted by $\{A, B, C...\}$. The precedences of logic operators (connectives) are parentheses, $\neg$, $\wedge$, $\vee$, $\longrightarrow$, $\longleftrightarrow$. Sometimes, we omit $\wedge$ for simplicity. Predicate calculus uses the universal quantifier $\forall$, the existential quantifier $\exists$, which stands for "there exists."

Bibliography

Ades, A.E. et al., (2006). Bayesian methods for evidence synthesis in cost-effectiveness analysis. *Pharmacoeconomics*, 24 (1): 1—19.

Aigner, M. and Ziegler, G.M. (2004). *Proofs from THE BOOK*. 3rd Ed. Springer, Berlin, Germany.

Turing, A.M. (1950). Computing machinery and intelligence. *Mind*, 59, 433—460.

Allesina, S. and Pascual, M. (2009). Googling Food Webs: Can an eigenvector measure species' importance for coextinctions? *Journal PLoS Computational Biology*. September 2009, Vol. 5, Issue 9.

Alonso, A. et al. (2006). A unifying approach for surrogate marker validation based on Prentice's criteria. *Statist. Med.* 25:205— 221.

Appignanesi, R. (2006). *Introducing Fractal Geometry*. Nigel Lesmoir-Gordon and Will Rood (2000). Icon Book Ltd. Cambridge, UK.

Astumian, R.D., and Bier, M. (1994), *Phys. Rev. Lett.*, 72, 1766.

Bar-Cohen, Y. (2006). *Biominetics*. CRC/Taylor & Francis. Boca Raton, FL.

Battersby, S. (2009). Statistics hint at fraud in Iranian election. *New Scientist* 24 June 2009.

Bell, J.S. (1964). On the Einstein—Podolsky—Rosen paradox. *Physics*, 1:195—200.

Benford, F. (1938). The law of anomalous numbers. Proc. Am. Phil. Soc. 78:551-572.

Benjamini, Y. and Hochberg, Y. (1995). Controlling the false discovery rate: A practical and powerful approach to multiple testing. *Journal of the Royal Statistical Society B*, 57(2):289—300.

Berezovskil, B. A. and Gnedin, A. V. (1984). *Optimal Choice Problems* (in Russian), Nauka, Moscow, Russia.

Berger, J. (2009). Adjustment for multiplicity, Subjective Bayes. Duke University Statistical and Applied Mathematical Sciences Institute. December 14—16, 2009.

Berger, J. O. and Wolpert, R. L. (1988). *The Likelihood Principle: A Review, Generalizations, and Statistical Implications*, 2nd ed. IMS, Hayward, CA.

Bernoulli, D. (1954). Originally published in 1738; translated by Dr. Lousie Sommer. (January 1954). Exposition of a new theory on the measurement of risk. *Econometrica (The Econometric Society)* 22 (1): 22—36.

Bernstein, S.N. (1928). *Theory of Probability*. Gostechizdat, Moscow, Leningrad. (In Russian; preliminary edition 1916.)

Beveridge, W. I. B. (1957). *The Art of Scientific Investigation*. Random House Trade Paperbacks; 3rd edition. New York, NY.

Bickel, P.J., Hammel, E.A., and O'Connell, J.W. (1975). Sex bias in graduate admissions: Data from Berkeley. Science 187 (4175): 398—404.

Birnbaum, A. (1962). On the fundations of statistical inference. J. Amer. Statist. Assoc. 57(298): 269—326.

Bishop, C.M. (2007). *Pattern Recognition and Machine Learning*. Springer, New York, NY.

Blachman, N.M. and Machol, R.E. (1987). Confidence intervals based on one more observation. *IEEE Transaction on Information Theory*. IT-33, 373—382.

Blackburn, S. and Simmons, Ke. (eds., 1999), *Truth*. Oxford University Press, Oxford, UK. Includes papers by James, Ramsey, Russell, Tarski, and more recent work.

Bollobás, B. (2001). *Random Graphs* (2nd ed.). Cambridge University Press. Cambridge, UK.

Bollobás, B. and Erdös, P. (1976). Cliques in random graphs. *Math. Proc. Cambridge Phil. Soc.* 80 (3): 419—427.

Boolos, G. (1989). A new proof of the Gödel incompleteness theorem. *Notices of the American Mathematical Society* 36: 388–90; 676. Reprinted in his (1998) Logic, Logic, and Logic. Harvard University Press, Cambridge, MA. 383—88.

Bradley, F.H. (1999). On truth and copying, in Blackburn et al. (eds.), *Truth*, 31—45.

Braess, D. and Über ein (1969). Paradoxon aus der Verkehrsplanung. *Unternehmensforschung* 12: 258—268.

Braess, D., Nagurney, A., and Wakolbinger, T. (2005). On a paradox of traffic planning (translated into English). *The Journal Transportation Science*, volume 39:446—450.

Bronshtein, I.N., Semendyayev, K.A., Musiol, G., and Muehlig, H. (2004). *Handbook of Mathematics*, Springer-Verlag Berlin Heidelberg, Germany.

Brown, J.R. (2000). *Thought Experiments*, in W.H. Newton-Smith, p. 529, John-Wiley. New York, NY.

Byers, W. (2007). *How Mathematicians Think*. Princetom University Press. Princeton, NJ.

Callaway, D.S., Newman, M. E. J., Strogatz, S. H., and Watts, D. J. (2000). Network robustness and fragility: percolation on random graphs. *Phys. Rev. Lett.* 85:5468—71.

Casella, G. and Berger, R.L. (2002). *Statistical Inference*, 2nd ed., Duxbury Press, Belmont, CA.

Chang, M. (2007). *Adaptive Design Theory and Implementation Using SAS and R*. Chapman & Hall/CRC, Taylor & Francis. Boca Raton. FL.

Chang, M. (2008). ABC of Bayesian approaches to drug development. *Pharm Med* 2008; 22 (3): 141—150.

Chang, M. (2010). *Monte Carlo Simulations for the Pharmaceutical Industry*. Taylor & Francis/CRC, Boca, Raton. FL.

Chang, M. (2011). *Modern Issues and Methods in Biostatistics*. Springer, New York.

Charig, C. R., Webb, D. R., Payne, S. R., and Wickham, O. E. (1986). Comparison of treatment of renal calculi by operative surgery, percutaneous nephrolithotomy, and extracorporeal shock wave lithotripsy. *Br Med J* (*Clin Res Ed*) 292 (6524):879—882.

Chisholm, R.A and Filotas, E. (2009). Critical slowing down as an indicator of transitions in two-species models. *Journal of Theoretical Biology*. Vol. 257, 1:142-149.

Chiu, S.S. and Larson, R.C. (2009). Bertrand's paradox revisited: More lessons about that ambiguous word, random. *Journal of Industrial and Systems Engineering* Vol. 3 (1):1—26.

Chow, Y., Robbins, H., and Siegmund, D. (1971). *Great Expectations: The Theory of Optimal Stopping*. Houghton Mifilin Co., Boston, MA.

Christakis, N. A. and Fowler, J. H. (2010). Social network sensors for early detection of contagious outbreaks, *PLoS ONE* 5 (9): e12948.

Clark, M. (2007). *Paradoxes from A to Z*. 2nd Edition. Routledge, Taylor & Francis Group. Boca Raton, FL.

Cohen, R., Erez, K., ben-Avraham, D., and Havlin, S. (2000). Resilience of the internet to random breakdowns. *Phys. Rev. Lett.* 85:4626—8.

Cohen, R., Erez, K., ben-Avraham, D., and Havlin, S. (2001). Breakdown of the Internet under Intentional Attack. *Phys. Rev. Lett.* 86:3682—5.

Cohen, R., Havlin, S., ben-Avraham, D. (2003). Efficient immunization strategies for computer networks and populations, *Phys. Rev. Lett.* 91: 247901.

Cox, D.R. (1958). *The Annals of Mathematical Statistics*, Vol. 29(2):357-372.

Cox, D.R. (2006). *Principles of Statistical Inference*. Cambridge University Press. Camridge. United Kington.

Cox, R.T. (2001). *Algebra of Probable Inference*. The Johns Hopkins University Press, Princeton, NJ.

Cox, M.T. and Raja, A. (2011). *Metareasoning: Thinking about Thinking*. The MIT Press. Boston, MA.

Crescenzo, A.D. (2007). A Parrondo paradox in reliability theory. *The Mathematical Scientist*, 32(1):17—22.

Dale, R. A. (2002). *Tao Te Ching*. 2nd edition. Barnes & Noble Books.

Davis, M. (1973). Hilbert's tenth problem is unsolvable. *American Mathematical Monthly* 80:233—269.

Darwin, C. (1859). *On the Origin of Species by Means of Natural Selection, or the Preservation of Favoured Races in the Struggle for Life* (1st ed.). London, the United Kingdom.

Dawid, A.P. and Mortera, J. (1996). Coherent analysis of forensic identification evidence. *Journal of the Royal Statistical Society*, Series B, 58,425—443.

de Finetti, B. (1974). *Theory of Probability* (2 vols.), John Wiley & Sons, New York.

Deckert, J., Myagkov, M., and Ordeshook, P.C. (2010). The Irrelevance of Benford's Law for Detecting Fraud in Elections, Caltech/MIT Voting Technology Project Working Paper No. 9

Dennett, D. (1995). Darwin's Dangerous Idea. Simon & Schuster. New York. p. 515—516.

Dickson, D.C. M. and Waters, H.R. (2002). The Distribution of the time to ruin in the classical risk model. ASTIN Bulletin, Vol. 32(2):299—313.

Dietrich, F. and List, C. (2008). A liberal paradox for judgment aggregation. *Social Choice and Welfare*, 31 (1):59-78.

Dmitrienko, A., Tamhane, A.C., and Bretz, F. (2010). *Multiple Testing Problems in Pharmaceutical Statistics*. Chapman & Hall/CRC, Taylor & Francis Group. Boca Raton, FL.

Dorigo, M. and Stützle, T. (2004). *Ant Colony Optimization*. MIT Press, Cambridge, MA, USA.

Duncan, R. L. (1967). An application of uniform distribution to the Fibonacci numbers. *The Fibonacci Quarterly*, 5:137—140.

Ebert, T., Merkle, W., and Vollmer, H. (2003). On the autoreducibility of random sequences. *SIAM J. Comput.* 32: 1542—1569.

Easley, D and Kleinberg, J. (2008). *Networks*, page 71. Cornell Store Press.

Easley, D. and Kleinberg, J. (2010). *Networks, crowds and Markets: Reasoning about Highly Connected World*. Cambridge University Press. New York, NY.

Edelman, D. (1990). A Confidence interval for the center of an unknown unimodal distribution based on a sample of size 1. *The American Statistician*. Vol. 44, No. 4: 285—287.

Efron, B. and Morris, C. (1977). Stein's paradox in statistics. *Scientific American*, 236 (5): 119—127. doi:10.1038/scientificamerican0577-119. http://www-stat.stanford.edu/~ckirby/brad/other/Article1977.pdf.

Einstein et al. (1953). Can quantum-mechanical description of physical reality be considered complete? Phys. Rev. 47, 777—780.

Erdős, P. and Rényi, A. (1960). On the evolution of random graphs. *Publications of the Mathematical Institute of the Hungarian Academy of Sciences*, 5: 17—61.

Erdős, P.; Rényi, A. (1959). On random graphs. I. *Publicationes Mathematicae* 6: 290—297.

Feferman, S. (2006). The Impact of the incompleteness theorems on mathematics. *Notices of the AMS*. 53:434—439.

Feige, U. (2004). You can leave your hat on (if you guess its color), Technical Report MCS04-03, Computer Science and Applied Mathematics, The Weizmann Institute of Science. New York.

Feld, S.L. (1991), Why your friends have more friends than you do. *American Journal of Sociology*, 96 (6): 1464—1477.

Finner, H. and Roter, M. (2001). On the false discovery rate and expected type I errors. *Biometrical Journal*, 43:985—1005.

Fisher, R.A (1935). The Logic of Inductive Inference. *Journal of the Royal Statistical Society*. Vol. 98, 1:39—82.

Fisher, R.A. (1922). On the mathematical foundations of theoretical statistics. *Philosophical Transactions of the Royal Society of London*. Series A 222: 309—368.

Flitney, A.P. and Abbott, D. (2002). Quantum version of the Monty Hall problem, *Physical Review A*, 65, Art. No. 062318.

Foreman, L.A; Smith, A.F.M. and Evett, I.W. (1997). Bayesian analysis of deoxyribonucleic acid profiling data in forensic identification applications (with discussion). *Journal of the Royal Statistical Society*, Series A, 160, 429-469.

Freeman, P. R.. (1983). The secretary problem and its extensions: A review. *Internal. Statist. Review*, 51 189—206.

Gadrich, T. and Ravid, R. (2008). The coupon collector problem in statistical quality control. *Proceedings of the World Congress on Engineering* 2008 Vol II. London, UK.

Gale, D. and Shapley, L. S. (1962). College admissions and the stability of marriage. *Amer. Math. Monthly* 69, 9—14.

Gamow, G. and Stern, M. (1958). *Puzzle Math.* New York: Viking.

Gardner, M. (1954). *The Second Scientific American Book of Mathematical Puzzles and Diversions.* Simon & Schuster. New York.

Gardner, M. (1970). Mathematical Games: The paradox of the nontransitive dice and the elusive principle of indifference. *Sci. Amer.* 223:110-114.

Gardner, M. (1986). Elevators. Ch. 10 in *Knotted Doughnuts and Other Mathematical Entertainments.* New York: W. H. Freeman, pp. 123—132.

Gardner, M. (2009), *Hexaflexagons, Probability, Paradoxes and the Tower of Hanoi.* Cambridge Universe Press, New York.

Gazzaniga, M.S. (2011). *Who's in Charge? Free Will and the Science of the Brain.* HarperCollins Publishers, New York.

Gints, H. (2009). *Game Theory Evolving.* 2nd Ed. Princeton University Press. NJ. p. 13—15.

Gorroochurn, P. (2012). *Classic Problems of Probability.* John Wiley & Son, New York.

Gödel, K. (1931), Über formal unentscheidbare Sätze der principia mathematica und verwandter systeme, I. *Monatshefte für Mathematik und Physik* 38: 173—98.

Godfrey-Smith, P. (2007). Conditions for evolution by natural selection. *Journal of Philosophy* 104:489—516.

Goh, K.I, Salvi, G. Kahng, B., and Kim, D. (2006). *Phys. Rev. Lett.* 96, 018701 (2006).

Green, E.D. and Elgersma, K.J. (2010). How often are ecologists wrong? Lindley's paradox in ecology. The 95th ESA Annual Meeting, PS48-179.

Gusfield, D. and Irving, R.W. (1989). *The Stable Marriage Problem: Structure and Algorithms.* MIT Press , Cambridge, MA.

Hacking, I. (1965). *Logic of Statistical Inference.* Cambridge: Cambridge University Press.

Haldane, J. B. S. (1945). On a method of estimating frequencies. *Biometrika*, Vol. 33, No. 3: 222—225.

Harmer, G.P. and Abbott, D. (1999a). Losing strategies can win by Parrondo's Paradox. *Nature*, 402:864.

Harmer, G.P. and Abbott, D. (1999b). Parrondo's paradox, *Statistical Science* 14:206—213

Harmer, G.P., Abbott, D., Taylor, P.G., and Parrondo, J.M.R. in *Proc. 2nd Int. Conf. Unsolved Problems of Noise and Fluctuations*, D. Abbott, and L. B. Kiss, eds, American Institute of Physics, 2000.

Harmer, G.P., Abbott, D., and Taylor, P.D. (2000). The Paradox of Parrondo's games. *Proc. Royal Society of London* A 456:1—13.

Hastie, T., Tibshirani, R., Friedman, J. (2001, 2nd Edition, 2009). *The Elements of Statistical Learning.* Springer, London, UK.

Havil, J. (2008). *Surpriseing by Solution to Counterintuitive Conundrums.* Princeton University Press, Princeton, NJ.

Haw, M. (2005). Einstein's random walk. *Physics World.* January. P.19—22.

Hellman, M. E. (1979). The mathematics of public-key cryptography, *Sci. Amer.* 241, 130—139.

Hemrick, E (1999). *The Promise of Virtue.* Ave Maria Press, Notre Dame, IN.

Henry, H.S. (January-March 2008). The Pesticide paradox. *Rice Today* (1): 32—33.

Herbranson, W.T. and Schroeder, J. (2010). Are birds smarter than mathematicians? Pigeons (Columba livia) perform optimally on a version of the Monty Hall dilemma. *Journal of Comparative Psychology.* Vol. 124(1):1—13

Hillel, M.B. and Falk, R.(1982). Some teasers concerning conditional probabilities. *Cognition,* 11 (2): 109—122.

Holyoak, K.J., and Thagard, P. (1995). *Mental Leaps: Analogy in Creative Thought.* MIT Press, . Cambridge, MA.

Hutchinson, G. E., (1961). The paradox of the plankton. *American Naturalist,* 95, 137—145.

Immorlica, N. (2005). Ph.D. Thesis. *Computing with Strategic Agents.* Massachusetts Institute of Technology 2005. Boston, MA, USA.

Iyengar, S.S. and Lepper, M.R. (2000). When choice is demotivating: Can one desire too much of a good thing? *Journal of Personality and Social Psychology,* Vol. 79 (6): 995—1006.

Jaynes, E.T. (2003). *Probability Theory: The Logic of Science.* Cambridge University Press, Cambridge, UK.

Jeanpierre, M. (2008). The Inspection paradox and whole-genome analysis. *Advances in Genetics,* Vol. 64:1—17.

Jeffreys, H. (1939). *Theory of Probability.* Oxford University Press. New York.

Jeffery, W.H. and Berger, J.O. (1991). Sharpening Ockham's razor on a Bayesian strop. Technical Report, #91-44C. Dept. of Statistics, Purdue University. Indiana, USA.

Johnson, N.L., Kotz, S. Kemp. A.W. (1993). *Univariate Discrete Distributions.* John-Wiley, New York.

Julious, S.A. and Mullee, M.A. (1994). Confounding and Simpson's paradox. *BMJ,* 309:1480—1.

Kamakura, W.A., Basuroy, S. et al. (2006). Is silence golden? An inquiry into the meaning of silence in professional product evaluations. *Quant Market Econ,* 4:119—141.

Kanazawa, S. (2009). Why your friends have more friends than you do, The Scientific fundamentalist: A look at the hard truths about human nature, *Psychology Today.*

Kerridge, D. (1963). Bounds for the frequency of misleading Bayes inferences. *Annals of Mathematical Statistics,* 34:1109—1110.

Key, E. S., Klosek, M.M. and Abbott, D. (2006). On Parrondo's paradox: how to construct unfair Games by composing fair games. *ANZIAM Journal,* 47(4):495—511.

Keynes, J.M. (1963). *A Treatise on Probability.* London: Macmillan.

Kim, J.S. et al., 2006, arXiv:cond-mat/0605324v1

Kleene, S.C. (1943). Recursive predicates and quantifiers, reprinted from *Transactions of the American Mathematical Society,* v. 53 n. 1, pp. 41—73 in Martin Davis 1965, *The Undecidable* (loc. cit.) pp. 255—287.

Knödel, W. (1969). *Graphentheoretische Methoden und ihre Anwendungen.* Springer-Verlag. New York. pp. 57—59.

Knuth, D. E. (1969). The Gamow-Stern elevator problem. *J. Recr. Math.* 2, 131—

137.

Kolata, G (1990-12-25). What if they closed 42d street and nobody noticed? *New York Times.*

Konopka, A.K. (2007). Systems Biology: Principles, Methods, and Concepts. CRC Press, Taylor & Francis Group, LLC. Boca Raton, FL.

Krulwich, R. (2009). http://www.npr.org/blogs/krulwich/2011/06/01/ 120587095 /ants-that-count, 11/25/2009.

Kuhn, T.S. (1996). *The Structure of Scientific Revolutions* . University Of Chicago Press; 3rd edition. Chicago, IL.

Lao-Tzu. (2002). *Tao Te Ching*, Tanslated by Ralph Alan Dale, Fall River Press, New York.

Lazer, D. (Ed. 2004). *DNA and the Criminal Justice System, The Technology of Justice.* MIT Press, Cambridge, MA.

Lehmann, E. L. and Romano, J.P. (2005). Generalizations of the familywise error rate. *Annals of Statistics*, 33:1138—1154.

Lester, P. J.; Thistlewood, H. M. A.; Harmsen, R. (1998). The Effects of refuge size and number on acarine predator-prey dynamics in a pesticide-disturbed apple orchard. *Journal of Applied Ecology,* 35 (2): 323-331

Leviatan, T. (2002). On the use of paradoxes in the teaching of probability. Tel Aviv University, Israel. ICOTS6.

Levin, M. (1984). What kind of explanation is truth? in Jarrett Leplin. *Scientific Realism.* Berkeley: University of California Press. pp. 124—1139

Li, D., Kosmidis, K. Bunde, A. and Havlin, S. (2011). Dimension of spatially embedded networks. *Nature Physics,* 7 (6):481.

Libet, B., Gleason, C.A., Wright, E.W., and Pearl, D.K. (1983). Time of conscious intention to act in relation to onset of cerebral activity (readiness-potential). The unconscious initiation of a freely voluntary act. *Brain*, 106 (3):623—642.

Lindley, D.V. (1957). A statistical paradox. *Biometrika*, 44 (1—2):187—192.

Lord, F. M. (1967). A paradox in the interpretation of group comparisons. *Psychological Bulletin*, 68 (5):304—305.

Macheras, P. and Iliadis, A. (2006). *Modeling in Biopharmaceutics, Pharmacokinetics and Pharmacodynamics: Homogeneous and Heterogeneous Approaches.* Springer. New York.

Magnasco, M. O., (1993). *Phys. Rev. Lett.*, 71, 1477.

Marques de Sá, J.P (2008). *Chance, the Life of Games & the Game of Life.* Springer-Verlag, Berlin Heidelberg, Germany.

Mates, B. (1981). *Skeptical Essays*, U. Chicago Press, Chicago.

May, R. (1972). Limit cycles in predator-prey communities. *Science,* Vol. 177, pp. 900—902.

Mayo, D. G. and Spanos A. (2010). Error and inference: Recent exchanges on experimental reasoning, reliability, and the objectivity and rationality of science. Cambridge University Press. New York.

McCarthy, J. (1953). *American Mathematical Monthly*, Vol. LX, No. 10, December 1953.

Mehra, J. (Author) and Rechenberg, H. (Contributor) (2000). *The Historical Development of Quantum Theory.* Springer. New York.

Miller, P. (2010). *The Smart Swarm.* The Penguim Group. New York.

Minor, D. (2003). Parrondo's paradox - Hope for losers! *The College Mathematics Journal*, 34(1):15—20

Mitchell, M. (2009). *Complexity, A guided Tour*. Oxford University Press. New York, NY.

Nagurney, A. (2010). The negation of the Braess paradox as demand increases: The wisdom of crowds in transportation networks. *EPL* (Europhysics Letters) Volume 91.

Nash, J. (1951). Non-coorperative games. *Annals of Mathematics*, 5:286-295.

Nehaniv, C.L. and Dautenhahn, K. (Eds., 2009). *Imitation and Social Learning in Robots, Humans and Animals: Behavioural, Social and Communicative Dimensions*. Cambridge University Press; Cambridge, UK.

Nickerson, R.S. (2004). *Cognition and Chance: The Psychology of Probabilistic Reasoning*. Psychology Press. NJ.

Nigrini, M.J. (May 1999). I've got your number. *Journal of Accountancy*. http://www.journalofaccountancy.com/Issues/1999/May/nigrini.

Parrondo, J. M. R. and Dini's, L. (2004). Brownian motion and gambling: from ratchets to paradoxical games. *Contemporary Physics*, volume 45 (2):147—157.

Petrovskii, S., Li, B.L. (2004). Horst Malchow, Transition to spatiotemporal chaos can resolve the paradox of enrichment, *Ecological Complexity*, 1:37—47

Pickover, C. (2001). *Wonders of Numbers*. Oxford University Press. New York, NY.

Pickover, C. (2006). *The Mobius Strip*. Thunder's Mouth Press, An Imprint of Avalon Publishing Group, inc. New York, NY.

Pickover, C. (2009). *The Math Book: From Pythagoras to the 57th Dimension, 250 Milestones in the History of Mathematics*. Sterling Publishing Company, Inc. p. 310.

Prior, A. (1961). On a family of paradoxes. *The Notre Dame Jounal of Formal logic*, 2:16—32.

Raimi, R.A. (1976). The first digit problem, *American Mathematical Monthly*, 83, number 7: 521—538.

Ramsey, F.P. (1927), Facts and propositions. *Aristotelian Society Supplementary*, Volume 7, 153—170.

Robert, C.P. (1994). *The Bayesian Choice: a Decision-Theoretic Motivation*. Springer-Verlag, New York, NY.

Robertson, B. and Vignaux, G.A. (1995) *Interpreting Evidence: Evaluating Forensic Science in the Courtroom*. John Wiley and Sons. Chichester.

Rosenzweig, L.M. (1971). The paradox of enrichment. *Science*, Vol. 171: pp. 385—387

Ross. S.M. (2003). Probability Models. 8th Ed. Academic Press. New York, NY.

Russell, B. (1908). Mathematical logic is based on the theory of types. *Amer. J. Math*. 30, 223.

Russell, S. and Norvig P. (2006). *Artificial Intelligence: A Modern Approach*, second edition. Prentice Hall, NJ.

Sackville Hamilton, H. (2008). The Pesticide paradox. *Rice Today* (1): 32—33.

Sagan, C. (1980). *Cosmos*, New York: Random House.

Sainsbury, R.M. (1995). *Paradoxes*. (2Ed). Cambridge University Press. New York, NY.

Salsburg, D. (2001). *The Lady Tasting Tea—How Statistics Revolutionized Science in the Twentieth Century.* W.H. Freeman and Company, New York, NY.

Sarkar, P. B. (1973). An observation on the significant digits of binomial coefficients and factorials, *Sankhya B*, 35, (1973), 363—364

Savage, R.P. (1994). Paradox of nontransitive dice. *The American Mathematical Monthly.* Vol. 101 (5).

Schelling, T. (2006). *Micro-Motives and Macro Behavior.* W.W.Norton & Company. New York, NY.

Scott, J.G. and Berger, J.O. (2010). Bayes and empirical-Bayes multiplicity adjustment in the variable-selection problem. *Ann. Statist.* Volume 38, Number 5, 2587—2619.

Searle, J. (1984). *Minds, Brains and Science*, Harvard Univ. Press, Cambridge, MA.

Seitlheko, M. (2010). The Birthday Paradox in Cryptology, Submitted in partial fulfillment of a postgraduate diploma at AIMS, May 2010.

Selvin, S. (1975a). On the Monty Hall problem (letter to the editor). *American Statistician,* 29 (3): 134

Selvin, S. (1975b). A problem in probability (letter to the editor), *American Statistician,* 29 (1):67

Seneta, E. (1993). Lewis Carroll's "Pillow Problem": On the 1993 Centenary. *Statistical Science*, Volume 8 (2):180—186.

Shackel, N. (2007). Bertrand's paradox and the principle of indifference. *Philosophy of Science*, 74:150—175.

Shamir, A. (1982). A polinomial time algorithm for breaking Merkle-Hellman cryptosystems, Research Announcement.

Shanker, O. (2007a). Defining dimension of a complex network. *Modern Physics Letters B* 21 (6):321—326.

Shanker, O. (2007b). Graph zeta function and dimension of complex network. *Modern Physics Letters B* 21 (11): 639—644.

Shubik, M. (1954). Does the fittest necessarily survive?, in Shubik, M. (ed.), *Readings in Game Theory and Political Behavior.* Doubleday, New York. p. 43—46.

Simmons, G. J. Cryptology (1979). The mathematics of secure communication, *The Math. Inte/ligencer.* I, 233—246, (1979).

Skiena, S. (1990). Stable Marriages. §6.4.4 in *Implementing Discrete Mathematics: Combinatorics and Graph Theory with Mathematica.* Reading, MA: Addison-Wesley, pp. 245—246.

Smullyan, R. (1978). *What is the Name of This Book?* . Prentice-Hall. Princeton, NJ.

Song, C., Havlin, S. & Makse, H. A. (2005). Self-similarity of complex networks. *Nature*, 433, 392—395.

Stein, C. (1956). Inadmissibility of the usual estimator for the mean of a multivariate distribution. *Proceedings of the Third Berkeley Symposium on Mathematical Statistics and Probability.* 1. pp. 197—206.

Steinberg, R. and Zangwill, W.I. (1983). The Prevalence of Braess's paradox. *Transportation Science*, Vol. 17 (3):301—318.

Storey, J., Taylor, J., and Siegmund, D. (2004). Strong control, conservative point estimation and simultaneous conservative consistency of false discovery rates: a unified approach. *Journal of the Royal Statistical Society*, Series B, 66(1):187—

205.

Stoyanov, J. (1997). *Counterexamples in Probability* (2th Ed.). John-Wiley. New York, NY.

Stutzer, M. (2010). A Simple Parrondo paradox, *The Mathematical Scientist*, Vol.35(1).

Székely, G.J. (1986). *Paradoxes in Probability Theory and Mathematical Statistics (Mathematics and its Applications)*. 1st Ed, Springer. New York.

Tang, N., Wu, Y., Ma, J., Wang, B., and Yu, R. (2010). Coffee consumption and risk of lung cancer: a meta-analysis. *Lung Cancer*. 2010 Jan; 67(1):17—22.

Taylor, H.M. and Karlin, S. (1998). *An Introduction to Stochastic Modeling*. Academic Press. New York. NY.

The *American Heritage Dictionary of the English Language*, Fourth Edition. 2000. http://www.thefreedictionary.com/dream

Tierney, J. (July 21, 1991), Behind Monty Hall's doors: Puzzle, debate and answer?, *The New York Times*.

Trybula, S. (1965). On the paradox of *n* random variables, *Zastos. Mat.* 8:143—154.

Tu, Y.K., Gunnell, D., and Gilthorpe, M.S. (2008). Simpson's paradox, Lord's paradox, and suppression effects are the same phenomenon—the reversal paradox. *Emerging Themes in Epidemiology,* 2008, 5:2

Turchine, v. (1991), www.pespmu.vub.ac.be/infinity.html

Vakhania, N. (2009). On a probability problem of Lewis Carroll. *Bulletin of the Georgian National Academy of Sciences*, vol.3 (3): 8—11.

Varian, H. (1972). Benford's law. *The American Statistician*, 26: 65.

Vidakovic, B. (2008). Bayesian Statistics Class handout. Brhttp://www2.isye.gatech.edu /~brani/isyebayes/handouts.html

von Ahn, L. (2008-10-02). Modeling network traffic using game theory. Science of the Web: Lecture 10. Carnegie Mellon University.

vos Savant, M. (1990). "Ask Marilyn" column, *Parade Magazine,* pp. 16 (9 September 1990)

Waibel, M, Floreano, D., and Keller, L. (2011). A quantitative test of Hamilton's rule for the evolution of altruism. *PLoS Biology*, 9(5). e1000615.

Wall, M.M., Boen, J., and Tweedie. R. (2001). *The American Statistician*. 55(2):102—105.

Walrand, J. (2004). Lecture notes on probability theory and random processes, University of California, Berkeley, pp.193.

Wang, D. and Bahai, A. (2006). *Clinical Trial: A Practical Guide to Design, Analysis, and Reporting*. Remedica, Medical Education and Publishing, London, UK.

Washington, L.C. (1981). Benford's law for Fibonacci and lucas numbers, *The Fibonacci Quarterly*, 19.2: 175—177.

Westfall, P.H. Tobias, R.D., Rom, D., Wolfinger, R.D., Hochberg, Y. (1999). *Multiple Comparisons and Multiple Tests Using SAS system*. SAS Institute. SAS Campus Drivew, Cary, North Carolina, USA.

Wikipedia.org; http://library.thinkquest.org/C005545/english/dream/function.htm

Willard, D.E. (2001), Self-verifying axiom systems, the incompleteness theorem and related reflection principles, *Journal of Symbolic Logic*, v. 66 (2):536—596.

Williams, M. (1996). *Encyclopedia of Philosophy, Supp.*, "Truth", Macmillan (Ed.).

p572—573.

Wilson, M. (November 2010), Using the friendship paradox to sample a social network, *Physics Today.*

Winson, J. (1985). Functions of dreams. library.thinkquest.org.

Wolpert, D. H., Benford, G. (2010). What does Newcomb's paradox teach us? Cornell University. http://arxiv.org/abs/1003.1343

Youn, H., Gastner, M.T., and Jeong, H (September 2008). Price of anarchy in transportation networks: efficiency and optimality. *Phys. Rev. Lett.* 101 (12): 128701.

Zucker, D.M., et al. (1999). Internal pilot studies II: comparison of various procedures. *Statistics in Medicine*, 19:901-911.

Zuckerman, E. W., Jost, J.T. (2001). What makes you think you're so popular? Self evaluation maintenance and the subjective side of the friendship paradox. *Social Psychology Quarterly*, 64 (3): 207—223.

Zwillinger, D. (2003). *CRC Standard Mathematical Tables and Formulae*, 31st Ed. CRC/Taylor & Francis. Boca Raton, FL.

http://www.beijingservice.com/attractions/summerpalace/xiequyuan.htm

http://designer-entrepreneurs.com/blog/illustrations/wedding_dress_jogger.gif

http://devrand.org/view/diceFinder

http://lucidapple.wordpress.com/2011/01/

http://math.mind-crafts.com/godels_incompleteness_theorems.php

http://mathforum.org/library/drmath/view/58492.html

http://mindyourdecisions.com/blog/2007/11/20/game-theory-tuesdays-a
-brain-teaser-and-related-trivia/

http://nobelprize.virtual.museum/nobel_prizes/economics/laureates/2002/
kahneman-autobio.html

http://plato.stanford.edu/entries/language-thought/

http://plato.stanford.edu/entries/qt-epr/

http://resources.aims.ac.za/archive/2009/mile.pdf

http://tuvalu.santafe.edu/events/workshops/images/e/e1/Chisholm_and_
filotas_2007_resilience_and_paradox_of_enrichment.pdf

http://www.doc.ic.ac.uk/~nd/surprise_96/journal/vol4/cs11/report.html

http://www.nap.edu/openbook.php?record_id=1575&page=69

http://www.plosone.org/article/info%3Adoi%2F10.1371%2Fjournal.pone. 0012948

Index

Q

T - #0500 - 101024 - C0 - 234/156/16 - PB - 9781466509863 - Gloss Lamination